CYCLES OF FIRE

*Z*eta Piscium (the star Zeta in the constellation of Pisces the Fish) is one of many examples of pairs of co-orbiting stars with different colors. Some of these pairs, called symbiotic stars, experience gas transfers from one star to the other. As gas is expelled from one star, gravitational forces in the system cause it to flow into a ring around the other. This affects the evolution of the target star, causing a different heating history than would be experienced by a single star of the same mass. The process is visualized here as seen from an imaginary planet orbiting the pair.

CYCLES OF FIRE

STARS, GALAXIES AND THE WONDER OF DEEP SPACE

WRITTEN BY WILLIAM K. HARTMANN
PAINTINGS BY WILLIAM K. HARTMANN AND RON MILLER
WITH PAMELA LEE AND TOM MILLER

WORKMAN PUBLISHING, NEW YORK

Text copyright © 1987 by William K. Hartmann
Paintings copyright © 1987 by William K. Hartmann, Ron Miller, Pamela Lee and Tom Miller

Library of Congress Cataloging in Publication Data

Hartmann, William K.
    Cycles of fire.
    Includes index.
    1. Stars.   2. Galaxies.   3. Astrophysics.
I. Miller, Ron   II. Title.
QB801.H26   1987   523.8   87-21778
ISBN 0-89480-510-X
ISBN 0-89480-502-9 (pbk.)

Book design by Charles Kreloff
with Mark Freiman

Workman Publishing
1 West 39 Street
New York, New York 10018

First Printing October 1987

10 9 8 7 6 5 4 3 2 1

Manufactured in Italy

ACKNOWLEDGMENTS

Special thanks to Alix Ott, who helped keep down the chaos level while the text was being prepared, and to James Abbot for photographs of a number of the paintings; to Judith Miller for being there; to Workman staffers Sally Kovalchick and Bob Gilbert for their help during editing, and to Charles Kreloff and Mark Freiman for creativity and patience during design of the book; and to Peter and Carolan Workman for their support and hospitality.

DEDICATION

To the memory of Nicolaus Copernicus, Charles Darwin, Harlow Shapley and Edwin Hubble, cosmologists all, in their own ways, who made us a part, not the center, of the universe

and

to the memory of Jules Verne, who convinced us that we could explore it.

CONTENTS

OUR GALAXY: THE MILKY WAY

A UNIVERSE OF GALAXIES

THE POSSIBILITIES OF THE UNIVERSE

CYCLES OF FIRE

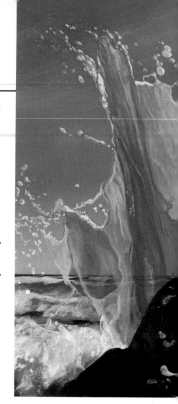

*P*lanets in multiple-star systems may be uncommon but could, if they exist, display many striking optical effects. This would be particularly true if they have atmospheres, as in this imaginary oceanic world. Atmospheric ice crystals, composed of different atmospheric gases—water ice crystals, carbon dioxide ice crystals, etc.—have different refractive properties and hence create haloes (rings of light around each star) and "sundogs" (bright concentrations of light along the haloes) of different angular size. The quadruple halo and sundog display, created by four suns shining through a sky of water-ice-crystal cirrus clouds, shimmers in an alien afternoon sky. This wide-angle view covers an angular width of 148° in order to show as much as possible of the halo display.

A WORD FROM THE AUTHOR

This is the third in our trilogy of illustrated books about the universe. In the first, *The Grand Tour,* we described the planets, moons and small interplanetary bodies circling our sun. In the second, *Out of the Cradle,* we explored what humanity may be able to *do* in this planetary system in the next fifty years or so. These two books were limited to the basic geography of the solar system. They covered only the tiniest fraction of the universe.

The universe beyond the outskirts of our solar system beckons to us. The dark sky at night—not the smog-shrouded, washed-out, streetlight-bleached sky of the city, but the *really* dark sky of the countryside, glittering with stars and misted with galaxies—enthralls us. Fortunately, even though we can't send our robots, brothers or sisters to them, we can still learn the secrets of the stars. We have telescopes. We have spectrographs that measure the light of each color. These instruments tell us which elements are present in the stars, galaxies and gas clouds.

Some of the telescopes are in traditional mountaintop observatories, where astronomers suffer through cold nights to get data. Other telescopes have been launched into Earth orbit, where they have a clearer view and can observe the kinds of light that don't make it through our atmosphere—X-rays, gamma rays, ultraviolet light and infrared light. Still other telescopes are giant radio antennas that receive a third kind of "light"—radio waves.

All these types of light, which are merely different forms of electromagnetic radiation, are messages from the stars. This book is about what we have learned from those messages.

The starry universe beyond the solar system gives new opportunities to painters as well as scientists. When we depicted the solar system in our earlier books, we knew a lot of specifics about the individual worlds: colors, types of surface materials, presence of clouds and rings, and so on. To help us, we had photos taken on the surface of the moon, Venus and Mars, as well as "aerial" photos of seven planets and more than a dozen moons.

In the case of distant star systems, there are no close-up photos. Astronomical research during the last century

told us the colors and sizes of the various types of stars, as well as the colors and shapes of gas clouds and galaxies. Research during the last decade revealed swarms of dust grains orbiting around many stars. Near a few stars, we have detected companion objects smaller than known stars but larger than Jupiter. Therefore, even though no actual extrasolar planet has been found at the time of this writing, most astronomers expect that at least some such planets exist.

The plausibility of planets near at least a few other stars is terribly provocative for an astronomical artist! It gives us a place to stand. It gives us a landscape. It gives us ... endless possibilities. There are only a finite number of sizable worlds in our solar system, but if there are planets around even one-half of one percent of all the stars, then there are a hundred million planets in our galaxy alone. Each offers its own potential for weird skies, sunsets, rock formations, volcanoes, Grand Canyons, Yellowstone Parks, fogs, fjords and fires, as well as for pterodactyls, platypuses, porpoises and people—or what passes for people on, let us say, Epsilon Eridani IV. Regardless of the likelihood of finding alien life, the possibilities for landscapes on far distant planets are enough to excite our imagination. In this book we try to be conservative about lifeforms (it's hard!), because they involve more speculation than planets do.

So lifeforms are set aside until book's end, but we give ourselves permission to imagine occasional planetary landscapes in the rest of the book. The recipe is three parts geological knowledge, two parts experience of light and landscape on our own planet (to keep us honest) and one part intuition. "Intuition" means all those bits of knowledge you've absorbed, but don't know you know, in your conscious mind. The possibilities for the entrancing, the awe-inspiring and the "sublime" (as the romantics and the American landscape painters of the nineteenth century used to say) seem limitless.

Landscapes on Earth have kept artists inspired for 500 years, and known landscapes in the solar system have offered us even further inspiration. But unknown landscapes of other stars and galaxies ...? What can we say? Only that we offer you here some realistic possibilities. We have constrained ourselves by known physical principles and have tried to act on known scientific findings. Thus our book depicts blue and orange and red stars; condensation of dust grains near them; aggregation of planets out of that dust; planetary heating, expansion, fracturing, volcanism, collision, cratering, erosion, cooling, contracting, glaciation, faulting; planets with and without atmospheres, with rings, with moons, with oceans of different condensed liquids (water isn't the only candidate). The universe contains far more real places than we know ...

PROLOGUE: THE GEOGRAPHY OF THE UNIVERSE

The night sky is dotted with stars. The ancients used them to navigate and to tell when to plant. They made constellations of them, pictures in the sky—the hunter, the lion, the bull, the greater and lesser bears. Once you get away from the glow and smog of the city, you'll see why the ancients made so much of the stars—the sky is full of them! Ancient people could actually see them better than most of us can, because their cities were dark at night. Over the dark streets of Babylon, Ur and Thebes hovered a thousand glittering stars and the hazy band of the Milky Way.

Go camping in the dark country, where the sky looks like space, and study the stars carefully. Look from one to another, and you'll see that they are not uniform points of light. They have different brightnesses, of course, but also they come in different colors. We have a sky full of gems.

What are they? During the Renaissance, in the 1500s, the charts of Copernicus and the telescope of Galileo revealed that a few of the gems are planets, like the Earth, moving in orbits around our own sun. But the vast majority of the diamonds that you see in the sky on any night are a million times bigger than planets, a thousand times hotter, and a million or more times farther away. Each of these gems is a sun, comparable to our own.

We live in a sea of sun-gems, each with its own distance, temperature, size, color. They are scattered through the sky like dust-motes floating in a beam of sunlight. The stars seem to twinkle in our sky, but the twinkling is caused by the light beam passing through miles of unsettled air, just as the image of a distant person shimmers when seen across a campfire whose heat disturbs the air. The light of these stars has traveled through space for

years and in many cases for centuries with virtually no distortion, only to be degraded into a shimmer in the last ten-thousandth of a second! The stars burn steadily in space, but if we watch them year in and year out, as astronomers have done for three centuries, we find that some of them flare up over periods of weeks or months—and that a few of them explode! We find, in other words, that stars have life cycles.

This book describes some of the varieties of stars, what makes them burn, how they are born and how they die, how they are distributed through space and whether they, like the sun, have little worlds moving around them—whether the sky is full of other Earths, as well as other suns.

The stars go on through the sky as far as we can see. The careful studies of astronomers have revealed that they are not scattered at random. They are clumped in groupings. The largest groupings, called *galaxies*, include a hundred billion stars at a time. Since the natures of galaxies were discovered only in the 1920s, we are virtually the first generation in the history of humanity to know that we live in giant systems of stars—stars that may, indeed, have other unseen Earths near them. No wonder our philosophies of the universe are so unsettled! No wonder wars still break out among different factions, with their different star gods, rituals, heavens and theories!

Galaxies of stars punctuate the sky as far as we can see. Indeed, the farthest, faintest objects we can see with the largest telescopes are distant galaxies, whose light has been traveling billions of years to reach us. In fact, we see them only in the form they had when their light left them: we see their early shapes, as they were billions of years

Ron Miller

From Earth to the cosmos: the scale of the universe. This series of figures zooms from our planetary home to the universe of galaxies. Each step is 10,000 times larger than the previous one. Step 1 (upper left) shows the moon circling Earth. Step 2 (lower left) shows the planets circling the sun (Pluto's orbit is inclined to the others). Step 3 (top middle) shows the sun and a nearby star. Zooming back to an even greater distance in step 4 (upper right), we see that all the local stars are only part of one spiral arm of our Milky Way galaxy. Step 5 (lower right) reveals that our galaxy is only one in a universe of millions of galaxies.

ago. It is in this way that astronomers gain insight not only about the most distant places but also about the earliest past. Our book, therefore, will talk not only about stars and galaxies but about the beginnings of things.

We have a word for the whole system of all things that exist, to the limits of our telescopes and beyond: *the universe*. The study of the universe is a study in hierarchies. There are three terms that many people distinguish only poorly. "Solar system" refers to the smallest system, our own backyard, the family of planets and other bodies that orbit around our sun. "Galaxy" refers to the middle-size systems, the clumpings of billions of suns. "Universe" refers to the largest possible system: all the galaxies—everything!

IMAGINING THE UNIVERSE

In order to conceptualize the universe of stars and galaxies, we need to imagine a scale model. Let's start with Earth. The features of the stellar universe are so big that we must make Earth very small in our scale model. Let's reduce the size of Earth so that it's bigger than most molecules but smaller than most dust-motes floating in a beam of sunlight in your room. Picture Earth as a microscopic dot, perhaps the size of a micro-particle of soot in a wisp of cigarette smoke. The sun is 100 times as big, more like one of the motes in the beam of sunlight; it measures about a hundredth of a millimeter across in our model.

The solar system is a set of microscopic motes circling this sun-mote in a disk-shaped system the size of a saucer. All the orbits of the planets out to Pluto would fit in the saucer. If you want to include the comets, which travel far beyond Pluto, imagine a swarm of atom-size mosquitoes buzzing around, most of them within a few house lots of the saucer.

The nearest star is another dust-mote floating one or two city blocks away. There are no dust-motes in between. Every one or two blocks across the city, we encounter another tiny grain of dust. These are the stars of our night sky. Some of the stars are giants—the size of sand grains! Some star-motes turn out to have other star-motes orbiting around them a few millimeters or a few inches away. Whether they have nonluminous micro-grains—planets—

moving around them is hard to tell from our vast distance of one or ten or a hundred blocks away.

Here and there throughout the city are puffs of smoke, as might be produced by a cigarette or a burning pile of leaves. These are clouds of gas and dust between the stars. Some of them may eventually spawn new stars, as we will see. They are called *nebulae*, from the Latin word for "cloud." They do not look like the clouds in our sky, which are often pushed by updrafts into cauliflower shapes; there is no "up" or "down" in interstellar space. These vast clouds, often glowing with different colors, are positioned sometimes raggedly, sometimes symmetrically, around a star that ejected the gas from which they were made.

If we back off far enough (a rocketship is needed, even to view our model!), we begin to discover that some two hundred billion of the glowing dust-motes are arranged in a vast disk the size of North America. Such a system of stars and dust is called a *galaxy*, from the Greek root for "milky," referring to its appearance as a milky glow. Our own galaxy is easy to detect with the naked eye. All the stars visible in our night sky are part of it, of course, but the more distant stars that are too far to be resolved as individual dots of light appear as a softly glowing band of light stretching across the sky—the Milky Way.

In our galaxy and many others, the stars are arranged in a vast, flattened disk. The reason we see the Milky Way as a band of light is that in our position on Earth we are looking edgewise out through the disk from a location inside the disk. We find ourselves encircled by a great band of light—the Milky Way—much as a nighttime tourist on a hill in Los Angeles might see the surrounding horizon as a glowing band of distant streetlights and houselights.

GIANT EGGS AND PINWHEELS: THE GALAXIES

In most galaxies, the stars are not scattered uniformly through the disk. There is a concentration toward the center. The most common types of galaxies are roughly egg-shaped concentrations. In addition, in many galaxies, stars are strung along spiral arms. These types of galaxies, from a distance, look like pinwheels. The arms are glowing streamers that curve out from a bright, crowded, central region. We live in one of these spiral arms, Earth being located in the outer regions of a spiral arm of our galaxy. Our sun, in the model, is a dust-mote in, say, Pittsburgh or Salt Lake City. Stretching over the state of Kansas is a great, dense concentration of stars and dust, a thick bulge in the otherwise phonograph-record-shaped system. In the center of this bulge is something poorly understood. It is smaller than a tennis ball in our model, or perhaps even microscopic, but it emits enormous amounts of energy. It may be some kind of superstar that occasionally explodes. It is called the *nucleus* of our galaxy. Most galaxies have such a central object; their mysterious, superenergetic properties are the subject of much current excitement among astronomers, as we will see later. In our model, some galaxies are the size of North America; some are bigger, comparable to Asia; and many others are only Australia-size or smaller. They are about as far apart on our scale as actual continents—several of their own diameters away from each other.

The universe is full of galaxies as far as the eye—or our telescopes—can see. In our scale model, this distance reaches at least as far as halfway across our solar system. The galaxies are not randomly distributed but are grouped into clusters in long, ragged strings or filaments, much as the leaves of an individual three-dimensional tree tend to lie in groups along branches. This stringy texture is thought to be a remnant of the universe's primordial structure, as we will see. If this model helps us grasp the scale of the universe, it also helps us realize the tremendous *range* in scale of phenomena that we will describe: from microscopic planets and stars to solar-system-scale distribution of galaxies. We can't make a single model in which we can see or sense everything, all at the same time.

In this book we start with the stars and the variety of star systems, and then expand our scale to encompass galaxies. We range back in time to see how the universe of galaxies may have started, and we discuss whether lifeforms may have evolved on the planets that we think may exist near other stars.

In each of these subjects, we encounter not only different objects but different time scales. We encounter billion-year cycles of evolution among stars, ten-year cycles of revolution of paired stars around each other, hundred-million-year cycles of rotation of ponderously turning galaxies, the 16-billion-year evolution of the universe since its start in a flash of light, and the possible cycles of evolution of planets, their climates, and their hypothetical lifeforms, guided by the evolution of their own star-suns. In various senses of the word, these are all cycles of fire.

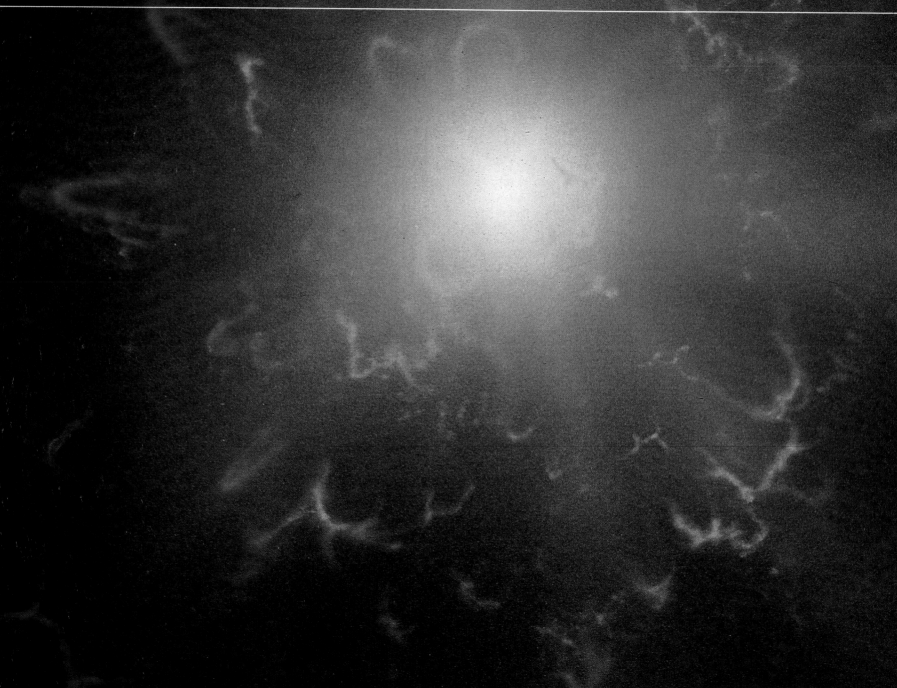

THE LIFE AND DEATH O

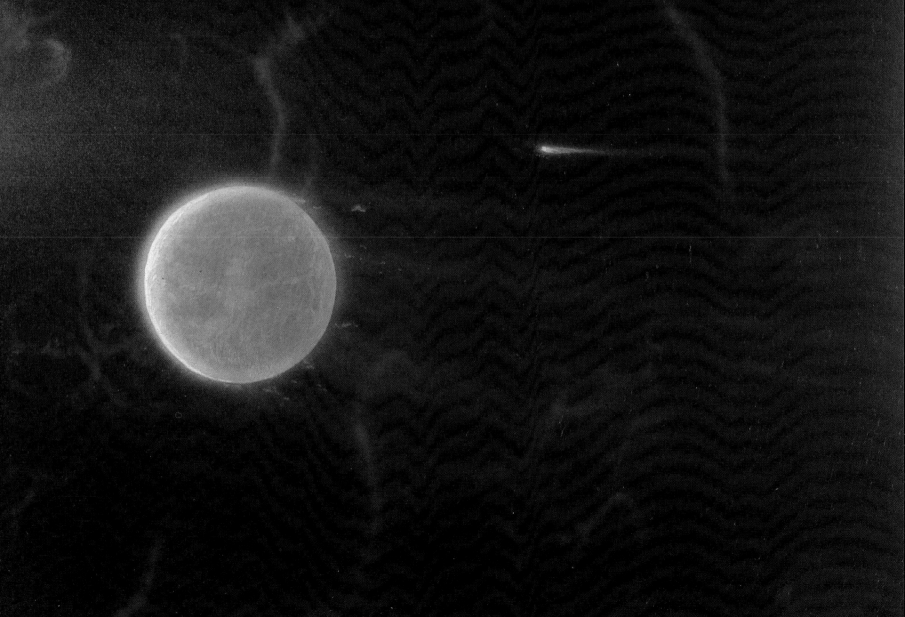

When a star explodes as a supernova, it blows off its outer layers of gas, creating an expanding nebula. Hydrogen from the outer layers forms filaments of gas glowing with the characteristic red color specific to this element. The explosion requires a period of a few days or less. The nearby planet is being turned into a giant comet as the outrushing radiation and gas strip away its atmosphere and volatile compounds. [Preceding page]

Ron Miller

HOW STARS WORK

Most people today know that stars are giant globes of incandescent gas. But what does this really mean? "Incandescent" means glowing, like a flame. But why should a giant ball of gas glow? What makes it "burn"? The gas that fills the room where you're sitting is not burning, and it would not burn if you lit a match to it.

The gas of a star "burns" only in the sense that it is so incredibly hot—thousands of degrees—that it gives off light somewhat the way the flame of a match or campfire gives off light. But the gas inside a star is not truly burning in the same sense as that of a match or a campfire. The heat in a match or candle flame, campfire or burning building comes from chemical reactions: oxygen in the air combines with atoms in the material in reactions that give off heat. "Combines" means the electron clouds forming the atoms' outer shells interact; when the clouds merge, atoms stick together and form molecules. The nucleus of each atom—the micro-kernel at the atom's center—is not

affected. In that sense, chemical reactions are only superficial, whereas nuclear reactions are profound. In nuclear reactions, the nuclei strike each other and may merge. Such reactions can produce vastly more energy than chemical reactions. The cloud of electrons around each atom's nucleus normally protects the nucleus from hitting another nucleus. In the centers of stars, however, the atoms are pressed together so hard that the nuclei hit each other. Thus, unlike the chemical heat of a burning match, a star's heat comes from nuclear reactions.

To understand how stars work, we need to ask how the atoms in a star's center can be pushed together hard enough to let the nuclei collide with each other. The answer is gravity, the force that makes all particles of matter in the universe attract all other particles. The more massive the particles, the greater the attraction. Also, the nearer the particles, the greater the attraction. So if you place enough atoms of gas close enough together, they will tend to fall toward each other. A giant cloud of gas, if dense

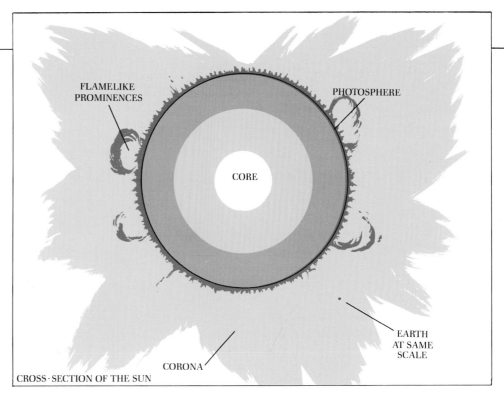

A cross-section of the sun shows the inner core in which nuclear reactions generate the sun's energy. This is surrounded by a thick zone of hydrogen and helium, which is incandescent but not hot enough to trigger nuclear reactions. The visible surface, called the photosphere, is the layer where the gas makes the transition between being opaque and being transparent; it is as insubstantial as the "surface" of a flame. An extended atmosphere surrounds the sun.

FLAMELIKE PROMINENCES

PHOTOSPHERE

CORE

EARTH AT SAME SCALE

CORONA

CROSS-SECTION OF THE SUN

enough, will start to contract. Gravity makes such a cloud contract; the interior heats up during the contraction, and the atoms in the center are pushed together hard enough to start nuclear reactions. The nuclear reations in the center heat the entire ball of gas, making it luminous; the gas cloud becomes a star. Thus, through gravity, masses of gas in the universe tend to turn into stars.

Let us call this our first principle of the universe, because we will use it to explain all stars and all the bizarre descendants of stars made famous by scientific discoveries of recent years: neutron stars, black holes and other curiosities of cosmic nature.

Here is our first principle, then: *gravity rules the universe.* Or, at least, gravity *tries* to rule the universe. Much of the stellar evolution that we witness in the universe—even the existence of stars themselves—can be viewed as a consequence of this principle. In fact, stars represent mere temporary disturbances in the successful reign of gravity.

THE MAGIC OF GRAVITY

If you think about it, gravity is a rather magical property. Astronauts can float out of a spaceship over to a crippled satellite and drift "weightlessly" around it. If the satellite were replaced by a small asteroid, say 100 yards across, the same would be true: the astronauts could "float" around it, poking among its craters. But when the asteroid gets big enough, a mile or so across, the astronauts floating nearby would notice themselves drifting toward the object and being pulled in by...what? The magical thing is that the pull happens even though the astronauts have never touched the asteroid and float at a moderate distance from it. Somehow, some "presence" of the asteroid reaches out to pull them toward it. This "presence" is called *gravitational force.* It is the same force that keeps the moon orbiting around the Earth, instead of drifting off into the solar system, and the Earth orbiting the sun.

In the 1600s, when Isaac Newton first recognized the existence of this force, naturalists were greatly troubled by its seemingly magical ability to affect objects without touching them—"action at a distance," it was called. Now scientists accept that this is simply a property of the universe and the matter in it. There are other, somewhat similar forces. Magnetism is one; a magnet can attract another magnet at a distance. But for the large-scale astronomical aspects of the universe, gravity usually rules over the other forces.

A gallery of hydrogen-burning stars, arranged in order of increasing mass, size and brightness. The tiny object at far left is technically not a star because it is so small (3 percent of the sun's mass, or about thirty times the mass of Jupiter) that it never gets hot enough inside to "burn" hydrogen in nuclear reactions. Called a substellar object, or brown dwarf, it glows by the heat generated during its formation. The smallest true star in the lineup is the object second from the left, with about 10 percent of the sun's mass; being cooler than the sun, it glows with a reddish light. Next comes a star somewhat larger, but still less massive than the sun. The fourth object is the sun, glowing with a yellowish white light. More massive stars (at right) are rarer but much brighter, hotter and bluer. The most common stars in the galaxy probably range in mass from about 0.3 to 1.0 solar mass.

Tom Miller

.03 .1 .3 1 3

The composition of the material does not matter when gravity is involved; gravity is "interested" only in how much material there is. The more mass in a parcel of material, the more strongly it attracts surrounding materials. In point of fact, most of the material in the universe is hydrogen gas—a situation that has great importance for the formation of stars. About three-fourths of the entire universe is hydrogen, and most of the other quarter is helium. The heavier elements—iron, carbon, oxygen, all the stuff we are made from—compose only a small percentage of the universe.

"Gravity rules" means that gravity is trying to pull all matter together. Thus, when we encounter a cloud of gas and dust in space, we know that all the atoms, molecules and dust grains in the cloud attract each other. If they're moving fast enough, they may disperse before they can fall together. But if they're moving slowly enough, gravity dominates, and they will inexorably find themselves falling together, clumping into a denser and denser cloud. Eventually, such a cloud can contract into a "small" ball, a few times the size of the sun, and this ball will become a star. In human terms, the contraction process is a slow shrinking,

taking thousands of years. However, this process is rapid in terms of astronomical time scales, and instead of calling it "contraction" or "shrinkage," astronomers call it "collapse" and speak of a cloud "collapsing" to form a star. It is as if, in our model, one of the clouds of cigarette smoke that represented a nebula suddenly became unstable and shrank down into a particle—a new dust-mote floating in the air.

These seemingly simple ideas contain some profound truths, at least when they're mathematically developed by physicists and astronomers. For instance, if we flesh out these ideas with numbers, we can learn why stars and galaxies exist—and how they were able to form in the first place! Suppose we could magically take all the atoms in the universe and spread them all uniformly so that there were no stars or galaxies, just thin, uniform gas. The individual atoms would be randomly moving, and because of this, even if all the atoms were initially evenly spread, they would eventually form local concentrations. Here, there would be "nodes" of stronger gravitational attraction. Ultimately, the magically smooth gas would break up into individual concentrations of diffuse clouds.

More complex physical calculations show that the

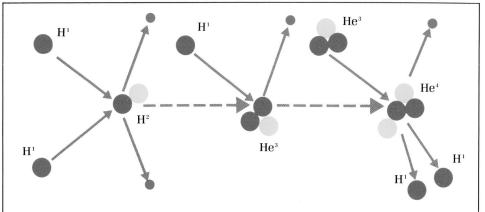

The structure of a complete atom. This sketch shows a helium atom whose nucleus includes two protons (black) and two neutrons (white). The nucleus is circled by two electrons. Inside a star, atomic particles strike each other so hard that the electrons are knocked off the atoms, leaving the nuclei free to hit each other, when pressures and temperatures become high enough.

A three-step example of fusion inside a star. In step 1 (left), two hydrogen nuclei (i.e., protons) collide and fuse, giving off a neutrino and a tiny positively charged particle called a positron; this produces a nucleus of "heavy hydrogen," designated H^2. In step 2 (center), the H^2 nucleus is hit by another hydrogen nucleus, giving off a photon of radiation and creating a nucleus of a type of helium designated He^3. In step 3 (right), the He^3 nucleus is hit by another He^3 nucleus, giving off two protons and a photon and leaving a nucleus of ordinary helium He^4. In this way, collisions among the lightweight hydrogen nuclei build up heavier nuclei such as helium nuclei, giving off energy at the same time.

NASA photo

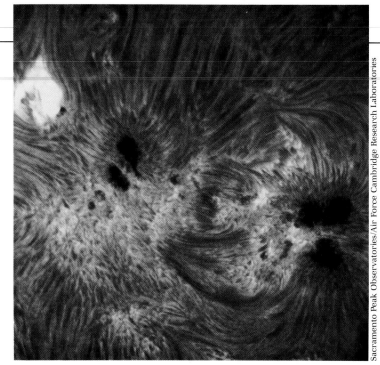

Sacramento Peak Observatories/Air Force Cambridge Research Laboratories

denser clouds would start to collapse. Large clouds, as they contract, would break apart into clusters of smaller clouds, each continuing to collapse on its own. As we will describe in more detail in a moment, the collapse of individual star-size clouds is temporarily stopped when the temperatures and pressures inside the cloud increase enough to start nuclear reactions. This is what forms a star. Without nuclear reactions to produce heat and outward pressure to fight gravity's reign, stars would collapse into strange, dense, cosmic bowling balls because there would be little to resist the compression process. In fact, when stars run out of fuel to sustain the nuclear reactions, they *do* collapse into the strangest bowling balls in the universe!

As long ago as 1926, the English astrophysicist Arthur Eddington pointed out that, by reasoning in this way and using adequate data from laboratory experiments, a physicist on a cloud-bound planet who had never seen the starry night sky could establish by calculation that the universe should be divided into clumps of matter made up of both galaxy-size and star-size masses. An imaginary physicist who lived below Jupiter's clouds, with adequate knowledge of gravity and atomic physics, could reason that the sky above his cloud tops should be filled with stars!

But why haven't all the clouds of gas and dust long ago contracted into stars? Why are there any clouds left? Many stars blow out new gas to resupply interstellar space with star-forming material for later generations of stars. In particular, very massive stars are unstable and explode, blowing out their own "used" gases to form new interstellar clouds. At the same time, these explosions compress nearby clouds with their resulting shock waves. A noncollapsing cloud, minding its own business, may suddenly be hit by an expanding shock wave from a nearby explosion that pushes its atoms closer together, thus increasing the gravity forces and inviting a collapse. There is a constant supply of gas recycling through many generations of stars in many galaxies.

Now it is time to describe in more detail how the collapse of a cloud stops, how the nuclear reactions "turn on"—in short, how a star is born. To answer these questions, we need a second fundamental principle of the universe drawn from subatomic physics. To state this principle, let's try to distill some complex science into something simple. So here is our second principle: *the movement of atoms is equivalent to heat; the faster the movement, the greater the heat and the greater the energy of the atoms.* This is an important principle. It will come back again and again to haunt us—or rather, to illumine us!

This 1972 photograph shows the surface of the sun not in ordinary light, but in just the red light given off by hydrogen atoms. Bright regions are disturbances where hydrogen is strongly emitting light; dark areas are cooler clouds emitting less light. [Far left]

This 1973 photograph shows a close-up view of a sunspot region on the sun's surface. Sunspots are cooler (hence darker) regions in which the sun's magnetic field is locally disturbed into swirl-like patterns. The hot gases' motions are controlled by the magnetic field pattern, giving the region its characteristic swirled appearance. [Near left]

Star probe. Our one opportunity to study a star at close range is provided by the sun. A proposal has been made to build a heavily insulated probe, here shown plunging into the sun's photosphere (at bottom). The probe would send data back as long as possible before being vaporized along with its instruments as it falls into the lower, hotter layers of the sun's atmosphere.

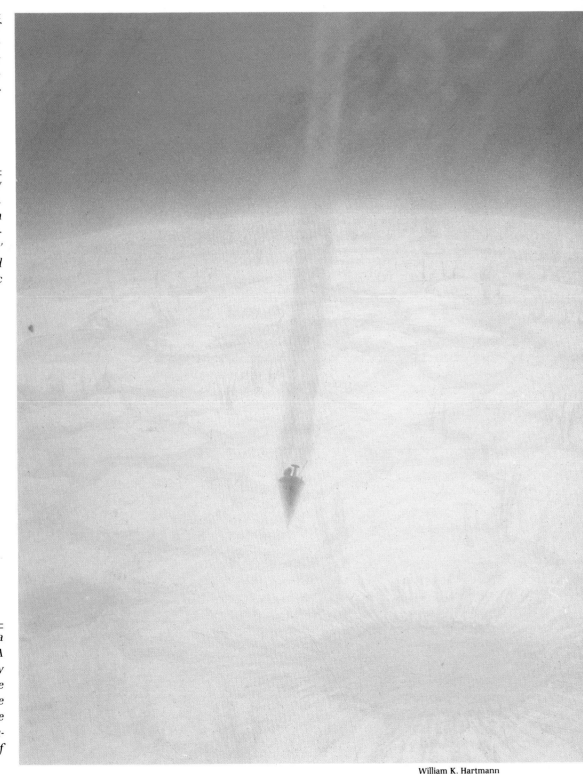

William K. Hartmann

HOW STARS FORM

Consider some hapless atom on the outskirts of a collapsing interstellar cloud. It feels the tug of gravity and starts falling toward the center of the cloud, accelerating as it falls. It will not fall all the way to the center of the cloud, because it will hit some other atom on the way. The impact will transfer some of its motion to the target atom. According to our second principle, this means that energy, i.e., heat, is gained by the atom as it falls and is then imparted to the target atom inside the cloud. In other words, *a cloud must become hotter as it contracts*, particularly in its center.

The chart on the facing page shows some of the things that happen to atoms as the temperature increases. As the cloud collapses, most of any solid grains of dust or ice trapped in the cloud will vaporize, or turn into gas, because the temperature soon reaches 2,000 K (3,140° F).* At such temperatures, the atoms in even rocky or metallic grains have enough motion, or energy, that the bonds that hold the atoms together are broken and the atoms are released one by one as gas.

As gravity continues to gain the upper hand, the atoms

National Optical Astronomical Observatories

The "Lagoon" Nebula (so-named after its appearance in small telescopes) is a region of star formation. This telescopic photograph shows masses of glowing dust, broken by clouds of dark dust. The densest dust clouds may be contracting toward formation of new star groups.

*A note on temperature: Scientists measure temperature in "Kelvin degrees," or, as they are called today, "Kelvins." These are abbreviated as "K," without the degree symbol. The Kelvin scale starts at absolute zero temperature. We will normally give equivalents in degrees Fahrenheit, or ° F. Recalling our second principle, from the end of the last section, it will make sense that absolute zero is the temperature at which atoms and molecules have no motion or energy. At 273 K (32° F) water molecules have enough motion, or energy, to break loose from the roughly crystalline structure of ice, causing the ice to melt into water. At 373 K (212° F) water molecules have enough motion, or energy, to zip loose from the surface of water, causing water to boil away into gaseous water vapor.

STATES OF MATTER IN THE UNIVERSE

TEMPERATURE (K) (F)	STATE OF MATTER		TYPICAL LOCATION	TYPICAL RADIATION EMITTED	TYPICAL VELOCITY OF PARTICLES*
	Nuclei breaking each other		Big Bang Black hole accretion disk	Gamma rays X-rays	
10,000,000,000 K 18 billion °F					16,000 km/s
	Nuclear fusion in ionized gas		Centers of stars	Ultraviolet light X-rays	
10,000,000 K 18 million °F					250 km/s
	Ionized gas		Outer layers and atmosphere of stars	Visible light	
5,000 K 8,500 °F					2 km/s
	Neutral gas		Earth's atmosphere	Infrared light	
500 K 440 °F					0.8 km/s
	Liquid		Bathtub	Radio waves	
300 K 80 °F					0.5 km/s
	Solid		This book	Radio waves	
0 K −460 °F					0 km/s

*in kilometers per second

Ron Miller

A protostar is forming in the heart of this black collapsing dust cloud. The cloud itself has contracted into a dust-rich globule silhouetted against the background. Inside the globule, the hidden protostar has just heated up to the point of becoming luminous; its dull red light filters out of the dark cloud. The protostar will get hotter and brighter until it turns into a full-fledged star, whose luminosity is driven by nuclear reactions in its own center.

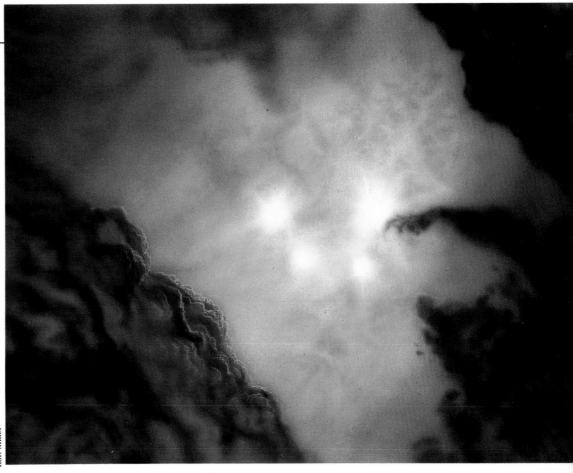

Ron Miller

At the heart of a star-forming system. In the glowing center of a dense, Orion-like nebula, a quadruple-star system is forming. Parts of the system are almost masked by clouds of dark dust that block the light of the system. Within 10,000 years, new stars may form and blaze forth within the system. Eventually, much of the nebular gas will be incorporated into new stars, and the rest will disperse into interstellar space.

fall farther in toward the center, the gas grows still hotter, and the atoms collide harder with each other. Since every atom consists of a nucleus surrounded by orbiting electrons, eventually the atoms collide hard enough to knock off the electrons. Such collisions begin to happen at temperatures of a few thousand Kelvins (close to 10,000° F). The gas is now said to be ionized because it is composed of charged particles: electrons (negatively charged) and free nuclei (positively charged). As the gas grows still hotter, these nuclei themselves begin to collide. Ordinarily, two nuclei will resist coming close together because the two positive charges repel each other, like the north poles of magnets. But once the temperature reaches perhaps ten to twenty million degrees (Kelvins or Fahrenheit) these collisions begin to happen.

Nuclei are strange little beasts, and when strange things happen they are crowded enough and move fast enough to hit each other. Fusion reactions, in which nuclei stick together, now occur and create the nucleus of a heavier element. For example, one of the earliest (coolest-temperature) fusion reactions occurs when hydrogen nuclei, the simplest of all nuclei (consisting of a single proton), collide. A series of such collisions leads to the creation of a helium nucleus (two protons and two neutrons stuck together).

The important thing to remember is that as light nuclei collide and fuse into heavier nuclei, *energy is released*. This is crucial to the nature of the universe. If it didn't happen, there would be no stars.

A STAR IS BORN

From the time the collapsing cloud becomes well defined until the nuclear reactions start, it is called a protostar. Afterwards, it is a star. The release of nuclear energy causes the inside of the protostar to become even hotter than it would have been from the simple heating due to

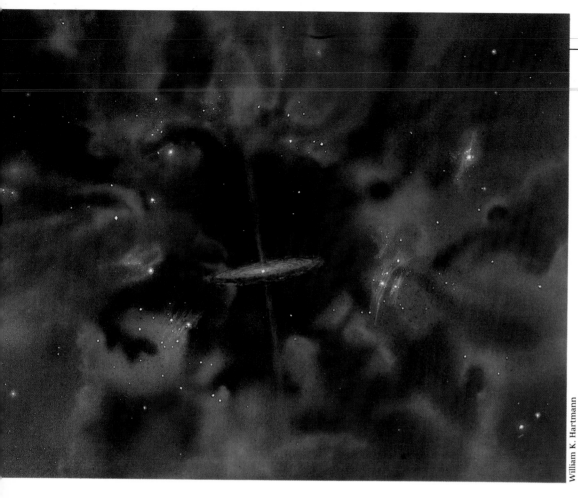

Near a dark molecular cloud (background, center) embedded in a colorfully glowing nebular region (background, edges), a star has just formed. It is surrounded and partially obscured by its disk-shaped cocoon nebula, some of which is being blown out in bipolar jets. The dust in the cocoon reddens the star's light much as atmospheric dust reddens our star at sunset.

collapse. The extra heat adds extra outward pressure to the gas, which is trying to expand. Thus, for the first time, there is an outward pressure resisting the collapse of the gas cloud, and the collapse stops. A steady state ensues and a star is born! A protostar, therefore, may last only some thousands of years—until it gets small and dense and hot enough to start its nuclear fires and create its own internal heat. It then stabilizes as it becomes a star.

In short, a star forms when a cloud of gas collapses until it gets hot enough to trigger nuclear reactions in the center. At this point, gravity is pulling inward but pressure is pushing outward, maintaining a balance and a constant size in the ball of hot gas. Inside our sun, for example, the central temperature is about 15 million K (27 million degrees F). In comparison, at the sun's surface, the temperature is only 5,700 K (9,800° F). The nuclear reactions in the sun, as in any star, are concentrated near the very hot center, in the region called the core.

Let us examine why a newly born, hydrogen-burning star is so stable. The fusion reaction rates are very sensitive to temperature. A slight increase in temperature produces an increase in the reaction rates and energy production, thereby raising the core temperature. This explains the stability of stars. Suppose that somehow you could take a hydrogen-burning star and expand it to a slightly larger size. The core would cool slightly. Gravity would begin to get the upper hand, making the gas atoms fall toward the center of the star. That is, the star would begin to shrink. The slightest contraction of the star, however, would drive up the central temperature. This in turn would strongly increase the reaction rate, thus adding to the outward pressure and bringing the star into a new equilibrium. Similarly, if you magically compressed a stable star, the reaction rate would go up so much in the center that the star would spring back to its original, stable shape.

Astronomers loosely say that the hydrogen in the star's

William K. Hartmann

Stars (and any planet systems they may have) form in loosely knit star clusters, usually associated with massive nebulae. Thus the sky of a newly formed planet (including that of primordial Earth) is likely to be dominated by the striking, eerie glow of star-strewn nebulosity. The moonlike new world shown here is illuminated by a massive blue-violet star, out of the picture to the right. Because of its high mass, the star is destined to explode in a few million years, blasting the surface of this planet with a lethal dose of radiation.

William K. Hartmann

William K. Hartmann

A panoramic photo of the winter evening sky shows the constellation of Orion and regions of "nearby" star formation. Several features, including prominent stars, are shown. The numbers beneath the stars' names indicate their distance from Earth in light-years. The Hyades and Pleiades are relatively young clusters, about 500 million and 100 million years old, respectively. The photo was made with an ordinary 35mm camera, 24mm lens, *f*2.8, 22-minute guided exposure on Tri-X film.

The "belt" and "sword" in the central part of the constellation of Orion. This is a region of current star formation. The middle "star" in the three-star sword, hanging from the belt, is not a star at all but the diffuse glow of the vast Orion nebula, a dust cloud in which a number of young stars have formed and in which various subclouds are probably collapsing to form future stars. The photo was made with an ordinary 35mm camera with a telephoto lens, *f*2.8, guided for 8 minutes, using 2475 Recording Film.

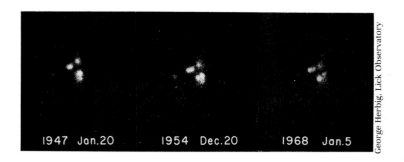

George Herbig, Lick Observatory

1947 Jan. 20   1954 Dec. 20   1968 Jan. 5

Changes in the heavens, related to star formation. These three photos show changes in a cluster of starlike objects over a period of twenty-one years. Once thought to be direct images of new star formation, these objects are now believed to be small clouds of gas and dust near newly formed stars; they are probably clouds of gas sporadically excited to a glowing state by collision with fast-moving gas ejected from the nearby new stars.

core is "burning," but this is not really the same as true burning. True burning is a chemical reaction—an exchange of electrons between atoms and molecules that remain intact during the reactions. In the nuclear "burning," atoms and molecules are not even present; they have been broken up into electrons and bare nuclei. It is the nuclei that interact in the fusion process. Nevertheless, both chemical burning and nuclear fusion involve a process of combining that gives off energy in the form of heat and light. So we will retain the astronomer's metaphorical language and speak of stars "burning" their nuclear hydrogen "fuel."

You will recall that *the universe is mostly hydrogen.* About 76 percent of the mass in the universe is in the form of hydrogen atoms. Almost all the rest is helium. As we noted earlier, only a small percentage of all matter in the universe consists of other elements such as carbon, oxygen and nitrogen (which help to make us) or of even heavier elements such as iron, aluminum and silicon (which help to make our planet). As a result of the great abundance of hydrogen, stars spend most of their lives "burning" this gas. "Normal stars"—most of the stars in the night sky—are of this hydrogen-"burning" type; i.e., they fuse hydrogen into helium and a few heavier elements.

Like an embryo, a star gestates rapidly inside its shrinking placental cloud and then is born to spend a relatively long, happy life as a youngster, adolescent and adult, living by consuming its hydrogen. As we will see, however, stars are just as subject to mortality as people are. Stellar death comes as the hydrogen runs out. At this stage, we encounter what might be called some strange terminal illnesses, leading to transfiguration . . .

STARS OF DIFFERENT MASS ARE STARS OF DIFFERENT TEMPERATURE

Mass is the most basic of a star's qualities. The mass of a star is simply the amount of material in it—the number of kilograms of gas. The total mass determines most of a newly formed star's future history, its size during most of its adult life, its rate of energy production and the duration of its lifetime.

We have commented that the gas and dust clouds inside galaxies tend to contract into star-size masses. As might be expected, however, there is some variation. For convenience, stellar masses are measured in terms of the sun's mass. A star with mass equal to that of the sun is said to have one solar mass. The most common stellar mass is about one-half to one solar mass. Most known stars fall in the mass range from one-tenth to ten solar masses. As we will see, objects much smaller never become stars, and objects much larger "burn" their fuel so fast that they explode.

If we line up a selection of hydrogen-burning stars in order of mass, we will find an array of different properties. The larger-mass, hydrogen-burning stars are generally somewhat larger in size, much hotter and much brighter than their smaller cousins. A ten-solar-mass, hydrogen-burning star is about five times as large as the sun, has three and one-half times its surface temperature (i.e., about 20,000 K or 35,000° F), and is 5,000 times as bright.

Conversely, a one-tenth-solar-mass, hydrogen-burning star is only 13 percent as big as the sun, has a little more than one-half the surface temperature (i.e., about 3,500 K or 5,800° F) and is only one-thousandth as bright.

Ron Miller

An imaginary world with an environment somewhere between that of Earth and Mars orbits at a relatively large distance from its double sun. The hot, bluish-white star is brighter than its cool, red companion.

Rigel is an extremely bright, "super giant," massive blue star. Its surface radiance is so great that it would dazzle the eye far more than the sun does. Here it is shown eclipsed by one of the moons of an imaginary planet, revealing its corona of hot gases. Wisps of red-glowing hydrogen gas are visible in the vicinity just outside the brightest area of the star's atmosphere. Rigel, the second brightest star in the constellation of Orion, is estimated to be 900 light-years from the solar system.

Ron Miller

These little stars are so faint that we see only the ones in the immediate neighborhood of the sun, while the bright, massive stars can be seen much of the way across the galaxy.

In visualizing how stars look, we now introduce a third important principle: *the hotter a star, the bluer the light it emits; the cooler a star, the redder its light.* This is an extension of what everyone already knows. As an electric stove's burner or a nail in a fire begins to heat up, it glows deep, dim red, then gets brighter and yellower. If it gets hot enough, we call it white-hot. If it got even hotter, we would perceive it as brilliant blue-hot.

Thus we can understand and visualize the great majority of known stars. They are spheroidal, glowing masses, composed mostly of hydrogen. The hydrogen in the hot hidden core is "burning" and the resultant radiation heats the entire star, causing the surface layers to glow with the dazzling light we see. Stars more massive than the sun are hotter and bluer; stars less massive are cooler and redder.

We used the word "spheroidal" to describe stars' shapes. Most stars are virtually spherical, like the sun. Stars rotate, however, and the rotation causes a slight flattening at the poles, just as Earth is flattened slightly by its own rotation. In the case of the sun, the rotation takes about twenty-seven days. A few stars rotate much faster, and the fast rotation causes a much flatter profile. A few such stars may be noticeably nonspherical, shaped more like eggs than oranges.

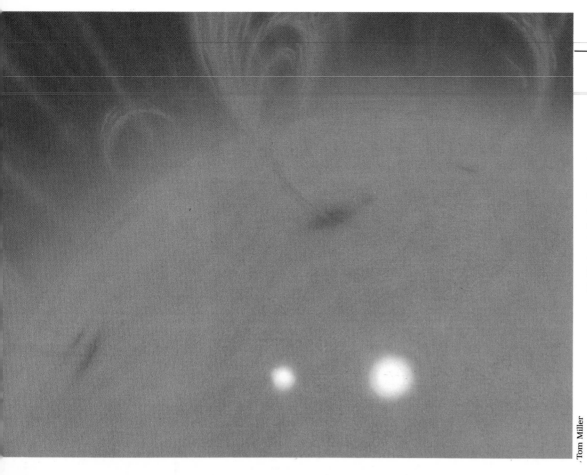

Relative sizes of a typical red giant and some hydrogen-burning stars. The red giant in the background has swollen to about 100 times the size of the sun after the hydrogen "fuel" in its core was exhausted. Its outer layers, cooler than the sun's and therefore redder, blow gas into interstellar space in massive prominences. In the bottom foreground are two familiar hydrogen-burning stars for comparison: the sun (yellowish white) and the more bluish white star Sirius, the brightest star in our nighttime sky. In terms of true luminosity, or total energy radiated, Sirius is 23 times brighter than our sun and the red giant is 10,000 times brighter than the sun.

HOW STARS EVOLVE: RED GIANTS

Clearly, the situation described in the last section cannot go on indefinitely. The core's hydrogen is being converted into helium. What if all the hydrogen in the core is turned into helium?

Remember our first principle: gravity is always trying to compress the star. As hydrogen runs out, a core of helium is being created. As the hydrogen-fusion energy generation rate tries to fall, gravity starts to contract the core, the temperature goes up . . . and a new cycle begins in the life of the star. The temperature rises so high and the helium nuclei become so energetic that the helium nuclei start to collide and fuse. Helium atoms fuse into heavier atoms, such as carbon, at a very fast rate. As a result, the core undertakes a very rapid energy production and undergoes consequent heating.

The extra heat causes the outer gas layers to expand rapidly into space, where they cool. The inner core is hotter, but the surface layers are cooler. Note that the star is now "burning" helium, not hydrogen, in its center. There may be a shell just outside the core that is still "burning" hydrogen. The outermost gas, having never been inside the core furnace, is still mostly hydrogen. This is the gas layer that we can see. Recalling our third principle, that cooler is redder, we can correctly predict that these stars will look red and very big.

They are called *red giants*.

In a sense, the "real" star is still about the same original size; that is, most of the mass is still buried deep in the center of the red giant, and the huge swollen atmosphere of red-glowing gas is no more substantial than a milk-

William K. Hartmann

Antares. *Above a rock-strewn plane on an imaginary planet, the cool red giant Antares fills much of the sky. This planet is as far from Antares as Saturn is from our sun. From this distance the sun would cover only a tiny angle of 1/20°, but giant Antares, which is probably bigger than the orbit of Mars, subtends an angle of almost 40°. A hot, blue star to the right orbits Antares in the distance, several hundred astronomical units away. The planet presents a geological mystery. If you were the leader of a team to analyze the geology from images from the first probe to land on this planet, how would you interpret the slant of all the boulders to the right? Are the rocks weathered out of a tilted layer? Has erosion by strong prevailing winds played a role? Such mysteries await all planetary explorers.*

William K. Hartmann

A star inside a star. A 1986 study by three Harvard-Smithsonian astronomers gave evidence for a companion star orbiting the famous red giant Betelgeuse, the star that forms one shoulder of the constellation Orion. Their data suggest that this close companion is in an elliptical orbit that dips inside the red-glowing, outermost atmosphere of Betelgeuse. In one interpretation (pictured here), the companion star could be a blue-white companion of a few solar masses. Such a system could form when the red giant expands and engulfs part of the orbit of the second star. The system would last only a short time, astronomically speaking, since drag forces will cause the second star eventually to plunge far into the giant and to merge with it.

weed's puffball around the seed. But this atmosphere is opaque and glowing red-hot, so a nearby observer cannot see the star-size core and instead perceives the star as a huge red object.

For another perspective, consider our own sun. It is a middle-aged star now. But someday it will become a red giant. When all the hydrogen has been exhausted from its core, the sun will expand and envelop Mercury, Venus and perhaps Earth in incandescent gas. The inner planets will be destroyed, probably compelled by the drag of the gas to spiral in through even hotter layers toward the sun's core until they vaporize! This will happen perhaps four billion years from now, according to astronomers' estimates, so we're safe for quite a while!

Red giants are stars entering old age. Because of their great rate of energy production, they are unusually bright stars, shining with 10,000 times the total light of the sun. For this reason, they can be seen a long way off through space. Many of the well-known stars in our night sky are distant red giants: Betelgeuse in the Orion constellation in the winter sky, for instance, and Antares in Scorpio in the summer sky. If you look at these stars and quickly glance

at other bright stars in the vicinity, you can usually see the redder color emitted by the giants' "cool" surfaces, where temperatures are "only" around 2,800 K (4,600° F). Betelgeuse is especially interesting because the blue-hot star Rigel is close by for comparison. Betelgeuse forms the left shoulder of Orion; the diagonally opposing star of the figure, making up the knee on the right side, is the 60,000 K (108,000° F) Rigel. If you glance back and forth between these two, you can easily see the color difference—or, in more physical terms, the *temperature difference.* Betelgeuse is cool; Rigel is hot.

Orion aside, you can glance around the sky on any night and see subtle color differences among the stars. Some are hot and blue, others are cool and rosy.*

---

*Note that the true physical colors associated with temperature are the *opposite* of the convention used by decorators. They call bluish colors "cool" and reddish colors "warm." This is presumably because bluish colors dominate on a cloudy, icy winter day (the sky light and light refracted through ice is bluish), while we think of a *red-*hot nail or the hot, reddish sun at sunset on a hot afternoon. But, of course, a red-hot nail is only the coolest light-emitting state; if the nail got hotter, it would give off bluer light.

William K. Hartmann

Both stars in this double-star, or binary, system have evolved into red giants, expanding until they touch. Such a system is called a "contact binary." The two stars are distorted in shape by their gravitational tidal interactions. The scene is also lit by a third red giant off-stage to the right. The landscape of this imaginary planet is deeply fissured due to changing tidal forces in the complex gravity field of the massive stars.

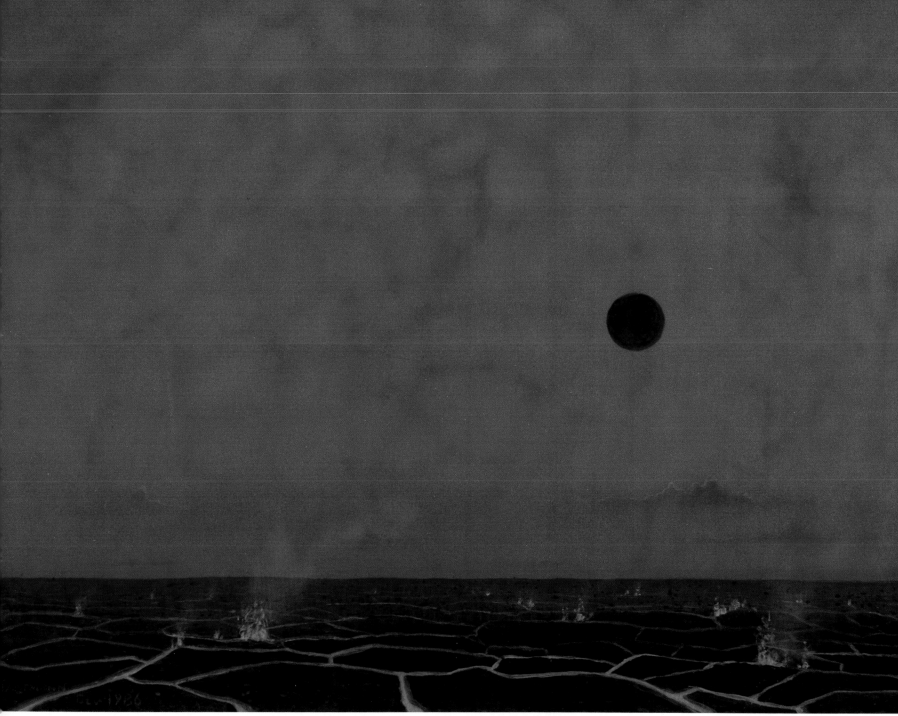

A dying world: A "red giant" sun has swelled to fill the sky of this planet. The increased heating of the planet by the now more luminous star has destroyed a formerly Earth-like climate, boiling off all the water and creating a crust of lava. The last gasps of volatile material create fumes from the steaming lava. While these fumes and vapors are temporarily added to the atmosphere, the atmosphere will eventually be blown off the planet by the radiation of the swelling giant. The planet's satellite makes a curious inkblot silhouette against the red-giant sky.

Let us clarify one semantic problem about stars. When we talk about stars being born, or being middle-aged, or entering old-age giant stages, we're referring to the life cycles of stars. But massive stars evolve faster than low-mass stars, in terms of actual years. Thus an "old-age" massive star may be much younger in years than an "old-age" low-mass star. Although this sometimes causes confusion, we are quite used to the same idea in terms of biology. Consider a whale, a human and a tortoise. The giant whale has a life-span of only about thirty years; the human, seventy years; the little tortoise, a century or more. If the three are born on the same day, the whale may reach middle age while the human is an adolescent; when the whale has reached old age, the human is middle-aged and the little tortoise is still relatively young. In this analogy, the whale is like a massive star, rapidly burning its fuel and evolving quickly; the human is like the sun; and the tortoise is like a low-mass star, burning its fuel very slowly.

FURTHER EVOLUTION: WHITE DWARF STARS

What happens when a star's helium fuel is exhausted? The rarer, heavier elements start to "burn"—and things become more interesting.

A star is a nuclear furnace heated by two energy sources. The ever-present background source is gravitational contraction, which gives the star its natal heat. The main source of energy during the star's active life is the burning of a series of nuclear fuels. First and by far the longest-lasting of these fuels is hydrogen. Hydrogen is followed by the brief outburst of helium. Then still shorter outbursts occur, involving a sequence of heavier elements such as carbon.

Whenever one fuel nears depletion, energy production starts to decline and so the outward-directed gas pressure

Ron Miller

David and Goliath. From the surface of an imaginary airless planet, we can see representatives of the largest and smallest stars combined in a binary star system. A white dwarf star orbits around a red giant, which is so distended and cool that it barely glows with a dull red color emitted from its diffuse outer gaseous atmosphere. The white dwarf, a collapsed star the size of Earth, lies between us and the giant.

Ron Miller

Delta Cephei: a star that changes size. About 1,000 light-years away in the constellation of Cepheus, the father of Andromeda, this inconspicuous star is the first known example of a class of stars that have reached an unstable point in their evolution, causing them to oscillate in size and brightness. Delta Cephei varies in size by about 8 percent, expanding and contracting in a period of five days, during which it changes in brightness by a factor of nearly two. Its peak brightness (near left) comes just after reaching the smallest size; its minimum brightness (far left) occurs just after reaching maximum size. The discovery of this behavior in Delta Cephei was made in 1784 by John Goodricke, a bright, young, deaf English astronomer who lived in the first generation in which the handicap of deafness was recognized as not being the same as mental retardation. Because the rate of pulsation is correlated with absolute brightness, astronomers can observe the pulsation rate of any star of this type, and use that to estimate the absolute brightness, and use that in turn to estimate the distance of the star. Thus these "cepheid variable stars" became the most important type of star in astronomy for measuring distances of star systems. At lower left is a bluish companion of Delta Cephei, at the unusually large distance of one-fifth light-year. [Left]

starts to wane. This allows gravity to begin its inexorable work: the star's core contracts and grows hotter; this in turn causes faster collisions among atomic nuclei in the core, prompting further fusion reactions. During the "burning" of the nuclei of each heavier element, neutrons or other particles are jammed into its nuclei and build up still heavier elements, some of which "burn" during the next cycle of reactions. Carbon fuses into oxygen, which in turn fuses into still heavier elements, and so on. Many reactions are involved.

In terms of cosmic time, all the post-hydrogen-"burning" activity happens very rapidly. The evolution from the giant stage to exhaustion of the last nuclear fuels happens within a small percentage of the total lifetime of the star. A star like the sun "burns" hydrogen for around 10 billion

Flare stars are stars that sporadically increase their brightness for periods as short as a few minutes. Most of these stars are believed to be cool, low-mass red stars. The flares, which make the star temporarily a few times brighter, are apparently outbursts of very hot gas that is much whiter or bluer than the color of the star itself. Probably the flares are giant examples of the much smaller flares that occur on our own sun, which are associated with magnetic disturbances and sunspots. [Right]

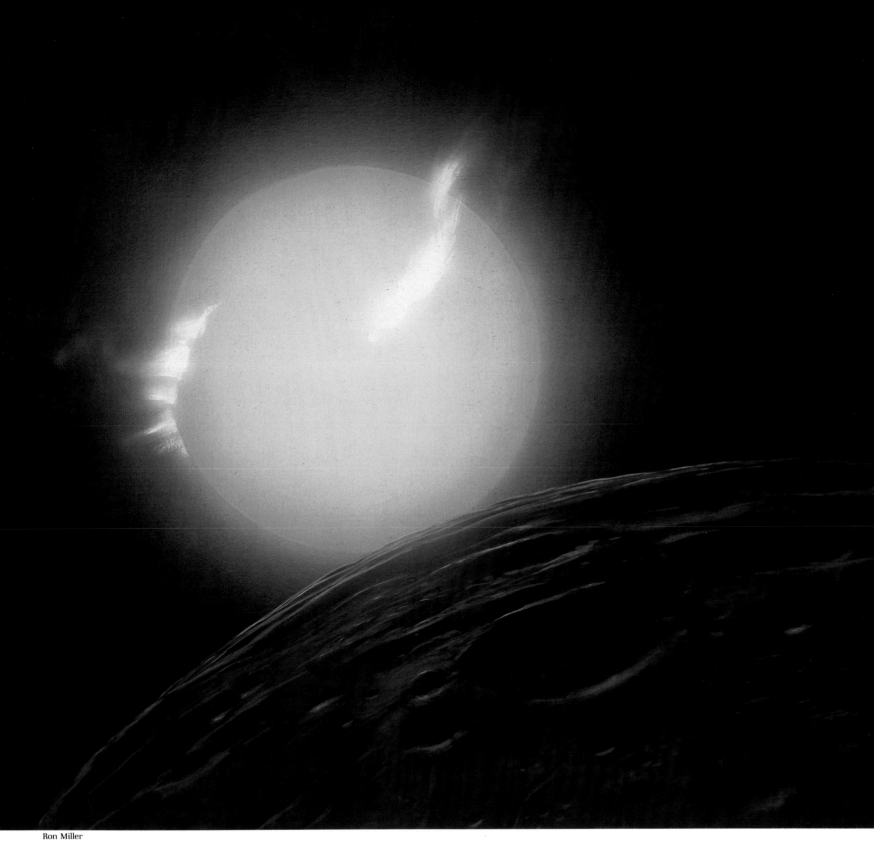

Ron Miller

The star known as "Wonderful." Astronomers around 1600 discovered that the relatively obscure star Omicron Ceti (Omicron in the constellation of Cetus the Whale) varies dramatically in brightness. It fades from view for much of the year but becomes a moderately bright star for a period of weeks. For this behavior, the star was named Mira (meaning "wonderful" in Latin). Here, Mira is seen from an imaginary planet as far away as Uranus is from our sun. The star is extremely expanded, about 400 times the size of the sun, and is apparently an unstable form of red giant. Its surface layers cool as it fades, dropping to temperatures about as low as recorded for any star: 1,600° C (2,900° F). At such temperatures, clouds of condensed material may form in the star's atmosphere, helping to dim its appearance. The snow fields of this planet would melt during the most luminous states reached by Mira in certain years. In the distance to the left is a bluish companion of Mira that was discovered in 1918.

years, but its final, most rapid changes may happen within a thousand years or less. These changes are also now sporadic. The core temperature rises and may trigger furious "burning" of some particular element. This in turn sends a burst of energy—heat-producing radiation—coursing through the star's outer layers.

One of the first indications that a red giant is continuing to evolve is that the thin, cool, red-glowing outer atmosphere gradually blows off into space. The remaining star, the inner core of the red giant, contains most of the mass of the original star, but is slowly shrinking in size and becoming hotter. Finally, the red giant reaches a point where many of the nuclei in its central core have been converted into iron nuclei. But iron atoms have a peculiar property: they resist further fusion reactions. They are the most stable of atoms in terms of nuclear structure. Therefore, iron atoms represent a final, cinder-like residue of nuclear "burning" inside stars. While humans may look for gold at the end of the rainbow, stars find iron at the end of the road!

Now that the star contains an exhausted, "unburnable" core, gravity has relatively free rein. A contraction begins in the star that will be unimpeded by new bursts of fusion energy. The phenomenon that ultimately stops this contraction is a strange property of subatomic particles. The electrons that were freed from the atoms during the star's formation formed a sort of electron soup in which all the nuclei floated. The strange property is that these electrons cannot be jammed together indefinitely even though gravity is pulling inward. One might think of them as something like marbles packed into a box, or atoms snapping into place in a crystal lattice during formation of a solid; only so many can fit per unit of space. When the electrons are unable to pack themselves any tighter, the collapse of the star halts.

The star has now reached a very old-age state. It is surprisingly small—comparable in size to a planet! Many such stars have diameters similar to that of Earth. Stars in this state might have been called "electron stars," because their size and structure are limited by electron packing; however, soon after they were discovered in 1915, they became known as "white dwarf stars," indicating their whitish or bluish-white color and small size, in contradistinction to red giants.

Here we can reemphasize the huge range of sizes displayed by stars. Middle-aged, hydrogen-burning stars range in size from only a little smaller than the sun to a little larger. But old-age stars vary from Earth-size (one percent the size of the sun) to the size of planetary orbits (100 times the size of the sun or more). We will see that a star's further collapse can produce even smaller stellar objects!

At this point, a note may be in order concerning the difference between size and mass. Remember that mass is the total amount of material in an object, usually measured by its weight. Even in everyday life we know that a group of objects with similar mass may have very different sizes. For example, a pillow that weighs a pound is much larger than a pound-size lump of lead. Similarly, a one-solar-mass hydrogen-burning star is much larger than a one-solar-mass white dwarf. Most of the stars we've discussed have about the same mass as that of the sun, but stars' sizes range from one percent to 100 times the size of the sun.

The final collapse that forms white dwarfs makes them very hot. Many have surface temperatures several times that of the sun, giving them a characteristic blue-white color. The joint Apollo-Soyuz orbital expedition of American and Soviet astronauts in 1975 discovered one of the hottest known stars, an obscure white dwarf with a temperature of 150,000 K (270,000° F)! These little stars are radiating away their heat without replacing it by new reactions, but they take an extremely long time to cool off. For this reason, white dwarfs are regarded as a relatively final state in the evolution of most stars. They are what happens in the long run when gravity defeats an ordinary star's attempt to stay active.

The thing that impressed everyone about white dwarfs when they were first discovered is their amazing density. White dwarfs' matter is jammed together so tightly that a thimbleful would weigh hundreds of tons. It would crush any desk on which it was placed!

Strange and dense as white dwarfs are, they are by no means the strangest or densest beasts in the stellar zoo—as we will see in the next section.

Some neutron stars are as small as asteroids. Here, somewhere in the galaxy, the path of a neutron star formed by the explosion of a supernova in a crowded young cluster has taken it through the asteroid-crowded cocoon nebula of a neighboring protostar—creating a bizarre stellar interloper among rocky worlds.

STRANGE CORPSES: NEUTRON STARS

We've been focusing on the most common stars—those with masses about one-tenth to ten solar masses. We've noted that the more massive the star, the faster it burns its fuel. Furthermore, the more massive the star, the stronger the force of its own gravity trying to make it collapse. We might suspect, therefore, that supermassive stars may undergo explosively rapid nuclear "burning" and experience supercollapses to very dense states. This turns out to be correct. And it leads us to the strangest of all star corpses: the *neutron star pulsar* and the *black hole.*

To get enough gravity to crush a star down to a diameter even less than that of a white dwarf, and a density even greater than a white dwarf's, we need more mass. A white dwarf configuration can survive only with mass less than 1.4 solar masses. If a "burnt-out" star has a mass more than 1.4 solar masses, its gravity will be great enough to crush the electron-dominated structure that holds up a white dwarf and it will continue collapsing to a still smaller size.

The next stable structure that can resist collapse is achieved by the network of free neutrons floating and jostling each other in the subatomic soup of electrons, protons, neutrons and other particles in the star. They are doing the real neutron dance. A point is reached where the neutrons resist being pushed any closer together. Nature won't allow more than a certain number of them per unit of volume, in a given physical state. They play the same role as the electrons in the white dwarf. This stable, exhausted-star configuration is called a *neutron star.*

A typical neutron star would measure only a mile or so in diameter! Here is a "star," or rather a stellar corpse, the size of an asteroid!

Of course, neutron stars are too far away for us to discern any details about their appearance. A few can be detected with large telescopes as faint points of light. But a neutron star, viewed from close range, would seem an incredible object. It contains twice the mass of the sun compressed to the size of an asteroid and is therefore amazingly dense. A thimbleful would weigh a hundred million tons—a density approaching that of the atomic nucleus itself!

Because its fuel is exhausted and the density is so high, the neutron star does not have the ordinary gaseous surface of a star. Some theorists believe it forms a crust of solid-like matter. Such matter would not be a solid in the familiar sense, but rather a bizarre dense substance of nuclear material. From a distance, it indeed might look like a weird asteroid, but a spaceship pilot would know something was wrong. The motions of the spacecraft would be deflected, pulled in, by the strong gravity of a two-solar-mass star. Gas or meteoritic particles in the neighborhood, perhaps thrown off by a companion star or a passing interstellar cloud, would crash onto the surface of the neutron star at speeds of thousands of kilometers per second, giving off bursts of light and heat.

Most neutron stars spin very fast. The reason for this can be understood if you visualize a ballerina doing spins. She starts with her arms out, but as she pulls her arms in, she spins faster. If any body, including a star, is set in spinning motion and its mass moves inward toward the center, its rate of spin will increase. A neutron star has pulled in its mass over a huge distance, from the million-kilometer diameter of an ordinary star to a size of only a few kilometers. As a result, neutron stars typically spin at a rate of several rotations per second, about as fast as a coin spun on a table! This would be another clue for the pilot of a passing spaceship: no ordinary asteroid could spin that fast, because its weaker rocky material would be torn apart by centrifugal forces at much slower rates.

An unusually perfect "bubble," blown off a star. When supernovae and other explosive stars blow off shells of gas, each shell expands spherically. Often it becomes distorted as it runs into neighboring gas clouds. The example in this photograph retains an uncommonly symmetric shape.

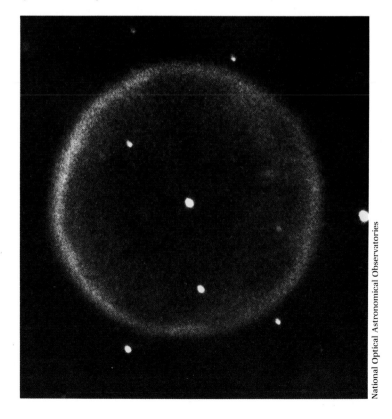

National Optical Astronomical Observatories

PULSARS: ALIEN BEACONS OR NEUTRON STARS?

At least some neutron stars have very strong magnetic fields near their surfaces. These magnetic fields derive from the star's original weaker, extended field being concentrated as it collapses into a neutron star. In such strong magnetic fields, ions—charged particles of in-falling gas—are drastically accelerated and give off beams of light. The light is emitted in certain directions and, as the star spins, the light beams play around the universe like beams from a spinning flashlight flung into the sky on a dark night. These flashing beacons have actually been photographed by astronomers.

Such objects are called *pulsars*, which is just another word for flashing neutron stars. When they were discovered in 1967, astronomers were astounded by the rapid, regular periodicity of the flashing. Some seriously thought that they might be artificially pulsing beacons put out along the spaceways by alien navigators! For a few months, until the neutron star theory was advanced, astronomers half jokingly called them "LGMs" (for "little green men"). But then theorists realized that their "LGMs" had just the properties expected for fast-rotating neutron stars.

Occasional tiny but abrupt changes in the rotation rates of neutron star pulsars have been observed. Theorists have suggested that these sudden changes in rotation are caused by sudden changes in the mass distribution in these pulsars. One theory is that a "starquake" occurs inside the pulsar, fracturing solid-like surface crust much in the way an earthquake affects Earth's surface layers. An-

William K. Hartmann

The exotic binary star SS 433, discovered in the late 1970s, has narrow jets of incandescent gas shooting out in opposing directions at about one-quarter the speed of light. The system probably contains an aging red giant (at left) which, through its expanding outer atmosphere, is expelling gas that flows onto a companion neutron star or black hole. The latter object is hidden in the midst of a so-called accretion disk of gas. The gas spirals inward and crashes onto the neutron star or black hole. Apparently, the gas in the jets has been squeezed out into the two polar directions as new gas falls in around the equator of the neutron star or black hole. The exact mechanism that forms the jets continues to mystify astrophysicists because acceleration to such enormous speeds is extraordinary.

Scene on an imaginary planet orbiting around a massive young star that has just exploded in a supernova outburst. In a matter of days the light and other forms of radiation will increase by a millionfold, blowing off the planet's atmosphere and melting rock on the daylight side. Flickering aurorae fluoresce in the sky as a flood of violent radiation from the dying star blasts the atmosphere of the doomed planet.

Ron Miller

other possibility might be the in-fall of some asteroid-like debris onto the surface.

HOW TO MAKE A NEUTRON STAR: SUPERNOVA EXPLOSIONS

How might a neutron star form? One could imagine magically adding mass to a white dwarf with, say, 1.3 solar masses. Once we get past the 1.4-solar-mass mark, the structure would become unstable and collapse into a neutron star. We wouldn't have to invoke magic; such a fate might befall a white dwarf if gas was shed onto it by a nearby star. In fact, this process seems to be happening right now in a bizarre star system called SS433, where a giant star is apparently dumping mass into a hot, luminous gas disk that probably houses a neutron star in its center.

But most neutron stars form during the explosion of a supermassive, unstable star. Remember, the more massive a star, the hotter its interior and the quicker it burns its fuel. Thus a star that is massive enough will go through its evolution with explosive violence. As the hydrogen is exhausted and other short-lived fuels come into play, the bursts of energy created by flashfires in the core literally blow the star's outer layers to smithereens. Such an explosion is called a *supernova*. The outer layers are blown away into interstellar space and the core, if it's more massive than 1.4 solar masses, collapses to form a neutron star.

The supernova explosion itself is one of the most spectacular events in all nature. It is a thermonuclear explosion involving a "bomb" the size of a star! The single exploding star rises in total brightness from the status of one insignificant star, in a galaxy of a billion stars, to a beacon whose total brightness matches that of all the other billion stars in the galaxy combined! The explosion takes several days, after which the star, as perceived at a great distance, slowly fades back to ordinary status during a period of weeks. Astronomers have watched such explosions in distant galaxies: a luminous point bursts forth, doubling the brightness of the whole galaxy, and then soon fades from view as the galaxy regains its former appearance.

In 1987, a supernova exploding in a nearby galaxy excited astronomers, becoming front-page news and the best-studied star explosion ever. Some aspects of its behavior confirmed long-held theories, while other aspects were unexpected. In particular, astronomers had correctly predicted the types and sequences of heavy nuclei blown out of supernovae: exactly that sequence was observed in the 1987 event! "It's rare in astronomy that theory gets as far ahead of observation as we were in this case," said one astrophysicist. Nonetheless, astronomers are still fine-tuning other aspects of their supernova theories as a result of the bonanza of new supernova data.

Most supernova explosions leave a stellar cinder, the exhausted inner core, which promptly collapses to form a neutron star. Neutron stars can form with masses anywhere from 1.4 to about six solar masses, although they all have small, asteroid-like sizes. What if a super-super-massive star explodes and leaves an exhausted core of, say, ten solar masses? If the remnant, exhausted star-core has more than about six solar masses, gravity is too strong for even a dense neutron star's structure to resist it. It collapses to an even smaller size than a neutron star. Gravity inexorably presses it toward the strangest density of all: *the black hole.*

he sizes of some small stars compared with our sun and Earth. In the background above the edge of the sun, flamelike "prominences" of glowing hydrogen rise into space. Shown to scale are Earth (silhouetted, dark blue) and a typical white dwarf star—the collapsed star-form produced by many stars when their core's hydrogen is exhausted. The two smaller black dots to the right represent the still more dense, contracted star-corpses of a neutron star and a black hole.

Tom Miller

CURIOUSER AND CURIOUSER: BLACK HOLES

As we follow the victory of gravity over exhausted star matter, things get, as Alice said, "curiouser and curiouser." This is because gravity compresses the material to ever greater densities, creating forms of matter unfamiliar to us in everyday life and gravity fields in which falling debris approach the speed of light. Most curious of all are the postulated structures known as *black holes.*

To explain black holes, let's start with *escape velocity.* This is the speed at which something must be catapulted to escape forever into space from the surface of an object (or from a specified position near an object) against the pull of gravity. If it is catapulted at less than escape velocity, and allowed to follow a free path against gravity's pull, it will simply fall back. To escape from a planet's surface typically requires a velocity of 4 to 60 kilometers per second (about 2 to 38 miles per second). To escape the sun, material has to be accelerated to approximately 600 km/sec (380 m.p.s.). To escape a white dwarf or a neutron star requires a speed an appreciable fraction of the speed of light, which is 300,000 km/sec (190,000 m.p.s.)! These high escape speeds are necessary because so much mass, with its

gravitational pull, is concentrated in such a small space.

Now imagine what might happen if even more mass were dumped onto a neutron star. The escape speed might be driven up, in principle, to more than the speed of light. *Photons*, the tiny, massless, particle-like entities that make up light itself, travel at the speed of light but no faster. In fact, a principle of modern physics is that *nothing* can travel faster than the speed of light. Thus we arrive at the conclusion that if you could create a massive enough, dense enough stellar corpse, nothing could ever be shot off it: not gas, not spaceships, not even light.

Astrophysicists believe that such objects—the so-called black holes—do exist. The name comes from the idea that anything that fell into one—a meteorite, spaceship or whatever—could never get back out because it would not have the energy to get up enough speed to escape.

Although black holes are among the latest darlings of astrophysics, the basic concept goes at least as far back as 1798! The French astronomer-mathematician Pierre Laplace realized then that a body might theoretically be so dense and massive that the escape velocity could exceed the speed of light. Such a body, he reasoned, could give off no light by which to be seen. (The speed of light had already been shown to be finite in 1675 by a Danish astronomer, Ole Roemer, who figured out a way to measure it by timing eclipses of Jupiter's moons.) Modern theories of black holes are considerably more complicated than Laplace's. They are being modified in particular by the subsequent discoveries of relativity and other effects. Nonetheless, a good way to think of black holes is to imagine a body so massive and dense that even radiation can't escape from the surface.

The accretion disk of a black hole is silhouetted against its companion "red giant," which fills the background sky. Some of the gas expelled by the red giant falls into the accretion disk and then spirals around, eventually crashing into the black hole at a very high temperature. Highest temperatures exist in the central part of the accretion disk because the material striking the disk at this point is moving at a fair fraction of the speed of light and imparting enormous energy to the disk. In some disks, material is apparently squeezed out from the center by magnetic forces along a line perpendicular to the disk plane; such a narrow, faintly glowing jet is seen emerging from the top and is barely visible behind the disk at the bottom. The black hole itself is nearly hidden in the disk's center. [Left]

The appearance of a black hole at close range is hard to predict. No light escapes from the object itself, but red-shifted light is emitted by extremely hot material falling into the object near its event horizon. Beyond the black hole (or rather, beyond the black hole's event horizon) is the bluish light emitted by the same hot material adjacent to the black hole and falling toward us on the far side. The outer regions of the accretion disks are cooler and less blue-white than the innermost regions. [Right]

William K. Hartmann

Astronomers think they've detected at least a few black holes in the universe. But how can this be if the black hole doesn't emit radiation that we can see? One of the astronomers' techniques is to observe the motion of another star that may be orbiting around the black hole. The velocity of the visible star and the size of its orbit allow astronomers to calculate the mass of the body around which it orbits. If they find that a star is orbiting around an invisible body with, say, ten solar masses, then they're likely to be dealing with a black hole.

The orbiting star does not get "sucked in" to the black hole because its orbital motion keeps it safely at a distance. The gravitational attraction of the black hole, while strong, is not fundamentally different from the gravity of any other object. It's a mistake to imagine a black hole "sucking in" nearby objects in a unique way. Just as Earth attracts the moon, or the sun attracts Earth, a black hole attracts a nearby star. And just as the moon orbits Earth, and Earth orbits the sun, the star can orbit the black hole.

Examples of such systems have been found, and the presence of black holes inferred. These systems must have formed as ordinary double stars, of which there are many. But the more massive star, the parent of the black hole, which had perhaps fifty solar masses, was unstable, evolved rapidly, exploded in a supernova explosion and left a core of perhaps ten solar masses that collapsed to form a black hole—while the less massive star evolved slowly, a sluggish witness to the extraordinary life of its companion.

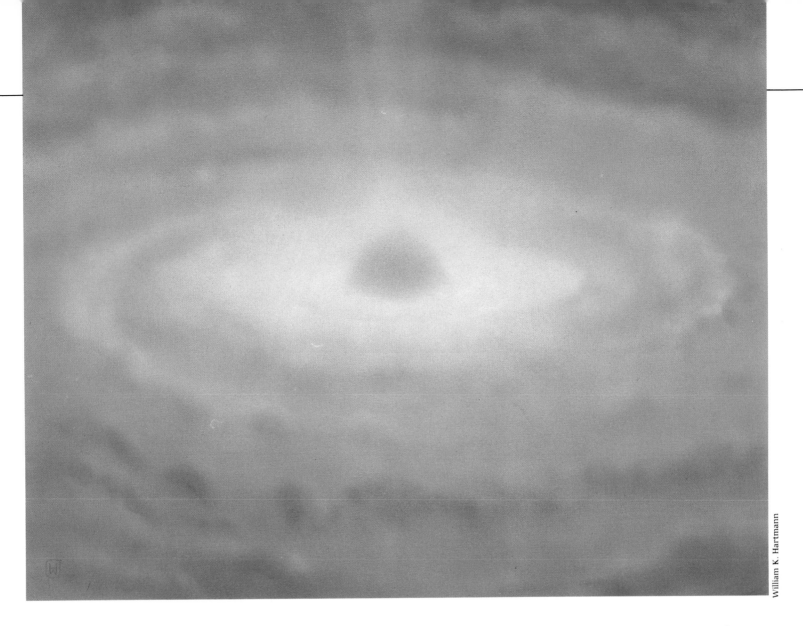

A second way to detect a black hole is by the gas (often from a companion star) that may be falling into it. For example, the companion star might be a red giant shedding its own gas. Falling toward the nearby black hole, the gas accelerates to tremendous speed—an appreciable fraction of the speed of light. The individual atoms are unlikely to fall radially onto the black hole without hitting each other; rather, their random motions lead to collisions. An orbital circulation pattern may result. In a roughly similar way, water does not readily flow straight into a drain but spirals around it. The disk of gas spiraling around a black hole (or any other star) and accreting onto it is called an *accretion disk*.

The extremely high-velocity collisions of the gas atoms in the accretion disk are equivalent to heat in the gas; hence, the gas falling into a black hole is heated to extremely high temperatures. The higher the temperature, the bluer and more energetic the radiation emitted by the material, so the gas falling into a black hole emits violently energetic photons of light: blue or ultraviolet light, even gamma rays and X-rays. Here you might want to say that this light cannot escape, but that statement holds true only within a distance where the escape velocity is equivalent to the speed of light. Such a distance is called the *event horizon*; we can detect no events inside it. Just outside this distance, however, radiation *can* escape. The outer parts of the accretion disk may be outside the event horizon. So, if a black hole with a hot accretion disk exists within the

Ron Miller

U *Geminorum: mass transfer in an eruptive variable system. The star U in the constellation of Gemini brightens by a factor of 100 in sporadic outbursts about 100 days apart. Studies reveal that the system is a close binary pair, with a redder star and a bluer star. The two stars are quite close together and have a very short orbital period, circling around each other every four and a half hours. The exact cause of the outbursts is uncertain; but as seen here from the frozen surface of an imaginary planet orbiting the pair, gas from the larger, redder star may form a ring around the bluer star, and the eruptions may occur as gases are drawn from the ring onto the surface of the hotter, bluer star. The system is sometimes called a miniature nova and is an estimated 300 light-years from our solar system.*

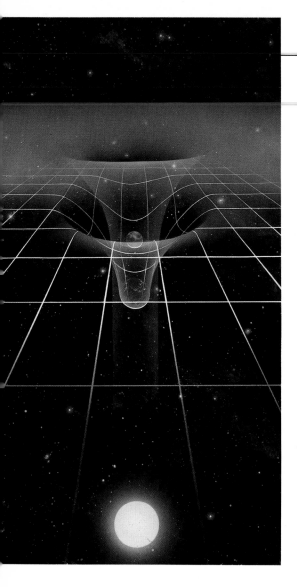

A schematic way of representing the presence of strong gravitational fields of massive objects in space. As explained in the text, the two-dimensional gridded surface represents space. A moving body such as a spaceship can be thought of as a marble rolling across this surface. In the foreground of the far-left view, a planet and its moon make a moderate depression in the surface, which could deflect the path of the marble. In the background, a massive star makes a much deeper "gravitational well." It would be hard for the marble to attain enough speed to climb out of the star's well against the star's gravity once it was caught in the lower part of this imaginary depression. The near-left view shows a representation of a still denser, more massive object. Because of its high density and mass, the object makes an extremely deep "dimple." A marble rolling into this pit would have a hard time climbing back out. Some mathematical researchers have suggested that in a black hole the dimple might "pinch out" at its end, leaving a detached "droplet," isolating the dense object from the rest of the universe!

Tom Miller

range of our instruments, we may detect it as an unusual source of X-rays and gamma rays.

In practice, both means of detecting black holes—studying orbital motion of a star and looking for gamma-ray or X-ray radiation—are combined. There are double-star systems in which a visible star is orbiting around an unseen, massive companion and the system is giving off gamma rays and X-rays. One example, a prime candidate for black-hole status, is called Cygnus X-1, the first X-ray source discovered in the constellation of Cygnus the Swan.

In addition to a handful of black-hole candidates in the form of massive, unseen companions to orbiting stars, there is another category of candidates. These are the so-called nuclei in galaxies. We will return to them later.

What are black holes really like? Since light is not emitted from or reflected off their surfaces, we cannot really form an image of the physical object itself. If we approached one, or went into orbit around it, we would *see* only its glowing cloud or accretion disk.

WORMHOLES?

Mathematical physicists, science-fiction writers, and moviemakers have been fascinated by a speculative possibility that extends the concept of black holes. This is sometimes called a *wormhole*. To understand it, we need to under-

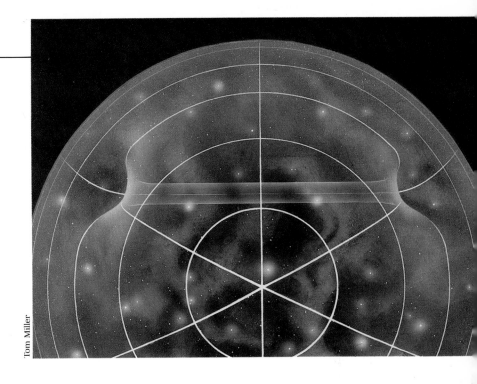

The "wormhole" concept. The distance across a curved universe might be "shorter" through the hypothetical wormhole than across "normal" space. Here we see a representation of curved space; the wormhole is a gravity-caused distortion that links two distant massive objects in a tunnel-like tube. (See text for further description.)

Tom Miller

stand that in relativistic physics an extremely great mass is conceptualized as warping space around itself. In order to grasp this concept, visualize a giant, stretched rubber sheet. Wherever we place a mass on it, a brick or a boulder, this mass distorts the sheet by making a depression. The sheet represents space, and each brick or boulder represents a star. A particularly massive, dense object—say, a lead cannonball—would make an especially deep, narrow depression that could represent a neutron star on the verge of becoming a black hole.

Now imagine a little, polished, steel marble that we can roll without friction across the surface of the sheet. It can represent a spaceship or an asteroid, or any object moving under the influence of gravity. If it rolled only across the smooth, horizontal parts of the sheet, the marble could go on forever. Even if it passed across the outskirts of the gravitational depression of a star, it would just dip into the edge of the depression, be deflected in direction and go on—like the Voyager spacecraft being deflected toward Neptune when it passed through the gravitational field of Uranus. If it approached too slowly into a deep gravity depression (the field of a massive object), it would roll down and hit the object. The marble would have to be propelled to "escape velocity" in order to roll back up the hill and out onto the surface of the sheet (interstellar space). So far, the analogy to stars and gravity is quite good. A black hole can be visualized as an infinitely deep, narrow depression in the sheet: no matter how fast we catapulted the marble back up the walls of the depression, it couldn't climb out of the pit; it would fall back.

One concept of a wormhole involves imagining a long, thin tube-like depression stretching down under the rubber sheet; instead of hanging straight down, it is bent and reattached to some other part of the rubber sheet. Our marble could roll into one gravitational depression and suddenly reemerge somewhere else in the universe from the other end of this "wormhole." A related concept involves imagining a black hole compressed to the size of a point, with infinite density. This could correspond simply to pinching off the end of a gravitational tube in our model so that a little teardrop-shaped volume of rubber could be enclosed and forever detached from the rest of our rubber-sheet universe. Some mathematical models of black holes have led theorists to suggest that these properties might actually exist in association with certain black holes.

But we must remember that mathematics itself is a tool of analogy. We use equations and numbers to make an idealized picture of the universe, which in its own way is as much an analogy as the rubber sheet, since minor physical effects are often ignored in even the most elegant mathematical theories. Therefore, our best picture of a black hole—which we believe, after all, to be a real, physical product of stellar evolution—is probably to imagine a solid-like object into which a spaceship could *crash* rather than a magical tunnel through which a spaceship could *pass*. Put it this way: at the present state of knowledge, I, for one, would not want to drive a spaceship into a black hole.

PLANETS AND STAR SY

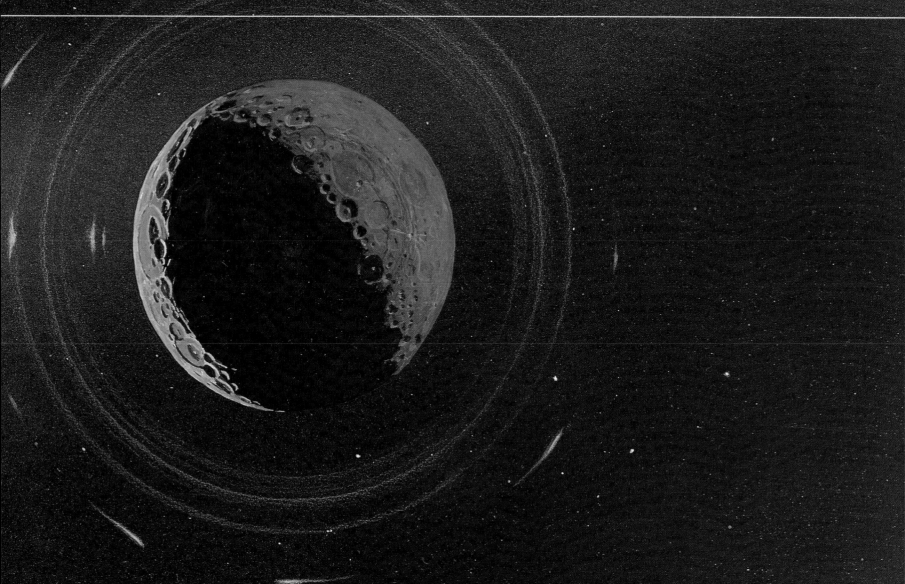

An alien ringed planet. In our solar system all ringed planets are gas giants, but there is no requirement that this be true; the break-up of a small body into a cloud of particles near enough to any planet could produce a system of rings around that planet. In this pole-on view from a position perpendicular to the plane of the rings, we see an airless ringed planet illuminated from the right and left by suns of different colors. The sun at left is eclipsed by one of the planet's moons. Nearby are two comets. Ringed systems also may have many different forms. Dynamicists, trying to explain the diffuse rings of Jupiter, the thick rings of Saturn, and the narrow rings of Uranus, have discovered principles that could produce rings of still different forms. For example, a French-American-Canadian team of dynamicists recently described dynamical processes that could produce arclike segments of rings controlled by the gravity of a small moonlet. [Preceding page]

William K. Hartmann

WHEN A STAR'S MASS IS TOO SMALL

We've been talking about ever more massive stars: sun-size stars that go through a normal hydrogen-burning evolution, and multi-sun-size stars that explode. But could a collapsing protostar have too little mass to make a star? Yes. A collapsing cloud of less than about 8 percent of a solar mass will never generate enough heat in its center to trigger nuclear reactions. Therefore, this is considered the minimum mass for a true star.

A collapsing cloud of somewhat less mass will produce what is sometimes called a substellar object—too big to be a planet and too small to be a true star. It may get quite hot by human standards; it may look like a small, cool star, but its heat is coming only from contraction and no stable, hydrogen-burning state is achieved. Remember, a true star initiates nuclear reactions in its core. The substellar object does not get hot enough to do this and therefore just keeps cooling, very slowly.

We have said that a substellar object is too big to be a planet. Since we want to compare such objects to planets, consider the largest known planet, Jupiter, as a standard of mass. Jupiter has about one-thousandth the mass of the sun. Thus, when we noted that the minimum mass for a true star is about 8 percent of a solar mass, we could have said that this is also about eighty Jupiter masses.

The surface temperature of such objects may be in the range of 1,000 to 2,000 K (1,340 to 3,140° F). At such temperatures the objects are red-hot. Seen at close range, they would not be dazzling, like stars, but instead would glow only with a dull, reddish light, perhaps obscured here and there by cooler clouds in their atmospheres. For this reason, such objects are often called "brown dwarfs"; their soft, brownish light is too faint to detect at great distances.

Astronomers are not sure how many of these curious substars may be scattered through interstellar space. In recent years, astronomy journals have been dotted with announcements of detections of several possible substellar objects, or brown dwarfs, of several Jupiter masses. However, as we will see, some of the reported detections have been controversial. Most of the reported brown dwarfs are orbiting around other stars, usually faint, low-mass stars. Some are separate stars (sometimes called "free-floaters").

Although these substellar objects are of great interest because of their possible link to planets, their names tend to be unfamiliar because they are associated with such faint, low-mass and obscure stars. Stars named Epsilon Eridani, Van Biesbroeck 8 and Gliese 623 are examples that have been reported to have companions of only a few Jupiter masses, according to work in the mid-1980s. Similarly, the isolated "star" LHS 2924 is cooler and less luminous than other known stars and may be a "free-floating" substellar object.

Astronomers hope to clarify the existence and statistics of substellar objects in the next few years. Space-borne infrared telescopes will be able to detect them much more readily than ordinary ground-based telescopes. Plausibly, future surveys may reveal one or more brown dwarfs closer to the solar system than the nearest star known today!

*A*t the end of its formative stage, the small red star in this scene, with only one-tenth the mass of the sun, is contracting and cooling. It is too cool and too faint to warm worlds at orbital distances similar to those of most planets in our solar system; however, during its formative collapse it was warmer and brighter for a brief period when a benign twilight existed on the imaginary foreground planet, as seen here. Now, as the star cools, the planet is fated to experience eternal winter.

Ron Miller

DOUBLE- AND MULTIPLE-STAR SYSTEMS

Among the gems twinkling overhead in the night sky, most seem to be isolated. Occasionally we see pairs, like Alcor and Mizar in the Big Dipper's handle. Are they really related to each other, or only paired by chance, like random couplings among a thousand salt grains spilled on black velvet? More specifically, are they really close to each other in three-dimensional space? Or are they merely lined up from our viewpoint, like the car in front of us and a car far up ahead as we look down an interstate highway?

If we set out to visit the stars on an imaginary hyper-space cruiser, we find that the majority of stars do not shine in solitary splendor. Most of the seemingly single stars in the sky actually have one or more fainter companions, which are either too immersed in the brighter star's glare or too faint to be seen easily from Earth with the naked eye. Compiling statistics on this from our Earth-based observations is hard, because many companions are too faint to detect among stars at large distances. This is particularly true of the low-mass stars just described. There are only a few dozen stars close enough for our observations to detect or rule out very faint companions. Nevertheless the statistics, as best we can tell, are something like this: if we flew our spaceship to a hundred random stars, we would find perhaps forty single-star systems, forty-five that are double stars, ten that are triple, three that are quadruple, and a couple that have five, six or more members in the system. Systems with more than two members are called *multiple-star systems.*

In other words, more than half of all "stars" are really star systems with two or more members. Our sun is normally classed as a one-star system, since there is no companion star orbiting our star—at least as far as we know now. Astronomers have occasionally searched unsuccessfully to see if there could be a faint, red, low-mass star orbiting the sun, far beyond Pluto. Yet we do know that the sun has companions. The largest is Jupiter, with one-thousandth the sun's mass; still smaller is Saturn, and so on. Thus the question remains open as to whether or not other, seemingly single stars may have low-mass companions that we cannot see. These could be low-mass stars, substellar objects, or planets.

Castor, the brightest "star" in the constellation of Gemini (the Twins), is actually an extraordinary system of six stars—three close pairs orbiting around each other. In this view, we are on an imaginary planet orbiting around the faintest and most outlying binary, a pair of low-mass red stars, Castor C, which is 1,000 astronomical units (12 solar system diameters) distant from the other stars. In the distance to the left are the other two pairs, Castor A and B, the brighter stars of the system. The whole system is located about forty-five light-years from our solar system and appears as a single star to the naked eye. An amateur telescope shows two "stars," which are actually the two pairs seen in the distance. It takes those two pairs about 400 years to orbit around each other; the foreground "red dwarf" pair requires some 10,000 years to orbit around the outskirts of that system. Spectroscopic studies revealed that each of the three seemingly single stars in the system is a close binary pair.

Ron Miller

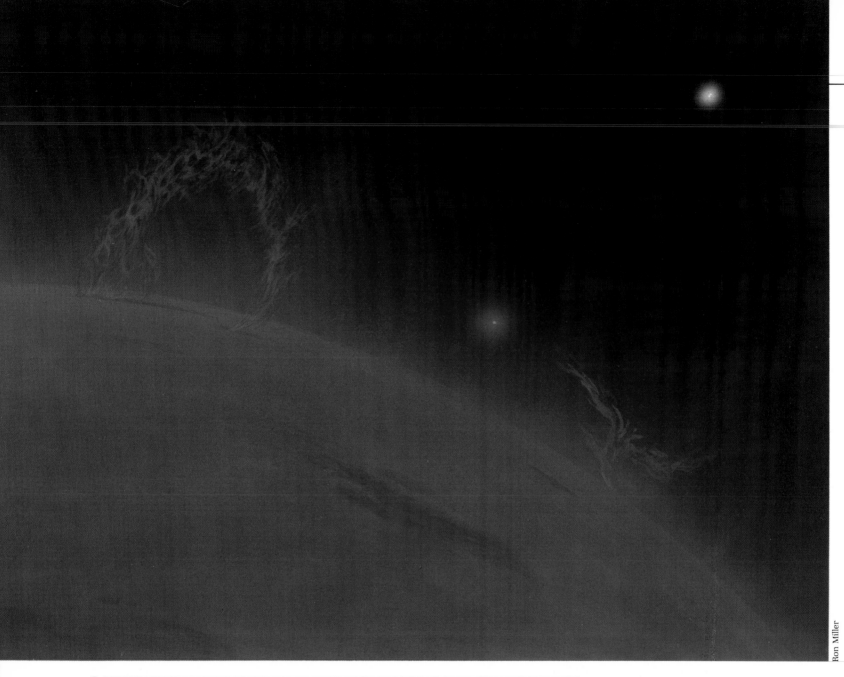

Capella, a well-known star in the northern constellation of Auriga, is an interesting system of four stars. Most of the light of the star we know as Capella is whitish light from a close pair of large stars, each about three times as massive as the sun. We view the system from a point in space close to a second pair of very faint red stars, which have masses and radii only a few tenths those of the sun. From a position above the surface of one of these stars (foreground), we can see the second, faint, red star (center right), estimated to be as far away as Pluto is from the sun. In the far distance (upper right) is the close pair of the system's brilliant white stars. At a distance of nearly two-tenths of a light-year, they are too close together to be distinguished easily with the naked eye, being only about as far apart as the Earth is from the sun. The unobtrusiveness of the red star in the center suggests the difficulties of detecting faint stellar or planetary companions if they are orbiting in systems of much brighter stars.

This makes for an interesting universe. There are all kinds of star systems. Consider just the double-star systems, or, as they are more properly called by astronomers, *binary star systems*. The brighter star, usually the more massive, is called the *primary*; the fainter one is called the *secondary*. Sometimes the two stars in a pair are very widely separated, with the companion farther from the primary than Pluto is from the sun. Sometimes the distance is intermediate, like that between Jupiter and the sun. Some companions are closer to their primary than Mercury is to the sun.

In some binary pairs, one star has turned into a red giant and is blowing off gas that then spirals through the system and crashes onto the secondary star. This is the apparent situation in the peculiar binary SS 433, mentioned earlier, where gas is spiraling from a giant star to what is probably a neutron companion. The strange thing in this system is that jets of gas are being squeezed out "above" and "below" the spiraling disk at a quarter of the speed of light. The mechanism that produces these jets is not understood.

Sometimes the secondary is so close that it is engulfed by the primary as the latter expands into a giant. Sometimes both stars have turned into giants and are so close

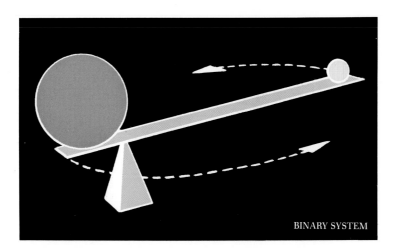

BINARY SYSTEM

I n a binary star system, the two stars orbit around their common center of gravity; thus, even if the smaller star cannot be seen, the motions of the larger one may reveal the presence of its smaller companion.

that their outer gas surfaces are actually touching! These are called *contact binaries*. They share their mass; that is, their gas may circulate from one object to the other through the contact region. A common atmosphere envelops them both. In these cases, the stars are usually distorted into teardrop shapes by tidal forces.

Some binary and multiple systems have stars of similar mass, while others have stars of quite different mass. In the latter case, the evolution rates are different. Some systems may have resulted when stars approached close to each other in a cluster and one (formed at a certain time) was captured into orbit around another (formed at a different time). Other systems formed together: their protostar cloud apparently broke up into two or more co-orbiting stars, perhaps because of the rapid spin of the collapsing cloud.

Since members of a single system can have different masses or evolutionary states, they may have different temperatures and, hence, distinctly different colors. We may have a yellow-white, sunlike star shining next to a red giant. We may have a reddish-orange star next to a hot, blue, massive companion. What an array of celestial jewels!

What an opportunity to conjure strange landscapes! Imagine standing on a moonlike planet lit by a red and blue binary pair hanging not far above the horizon in the late afternoon sky. Mountains and rock spires cast two shadows, one from each star. Where the shadows overlap, no light strikes and the shadows reach their darkest depth. But light from the red star reaches part of the shadow cast by the blue star. No blue light reaches such a region, because it is in a shadow cast by the blue star, so that part of the shaded region appears red. Meanwhile, part of the shadow cast by the red star is illuminated by the blue star, and that part of each shadow appears blue. Thus we would have a landscape with bicolor shadows picking up the tones of the two suns.

To our knowledge, this eerie effect was described first not by practicing astronomers, who have less interest in things' appearances than in their physics and chemistry, but by astronomical writers and painters. The French astronomical popularizer Camille Flammarion, in his 1894 compendium of astronomy (republished in English in 1907 as *Popular Astronomy*), talked about double stars of differ-

Ron Miller

W *Ursa Majoris. The star W in the constellation Ursa Major is the prototype of a
famous class of close binaries. Two whitish
stars, slightly more massive and slightly larger
than the sun, orbit around each other so
closely that their outer atmospheres touch. In
contrast to the twenty-five days required for
the sun to rotate, these two stars travel entirely
around each other every eight hours. Their
centers are only about one million miles apart.
Due to the close distance, gravity distorts the
stars into egg-shaped ovals. It is possible that
such close pairs form when rapidly rotating
protostars split into two lobes during their
prestellar collapse. W Ursa Majoris is an estimated 200 light-years from the solar system.
The star is viewed from an imaginary planet
that might be orbiting at a large distance from
the pair.*

Ron Miller

T he Flammarion effect on a ringed planet. This Saturn-like world orbits around an offstage
binary star pair with red and blue colors. The red star lies slightly above the plane of the
rings (here seen edge-on); the blue star lies below the ring plane. Only the red star's light falls
into the ring shadow produced by the blue star, coloring it red. Similarly, in the area of the ring
shadow cast by the red star, only blue light falls, coloring that shadow blue.

ent colors and how they could bathe different sides of their planets in contrasting hues of light. Lucien Rudaux, a French astronomical painter, writer and active amateur astronomer with his own observatory, subsequently described the effect of multicolor shadows in planetary landscapes (e.g., in *The American Weekly,* October 4, 1936). Rudaux, the grandfather of astronomical art, was succeeded by the father of astronomical art, Chesley Bonestell, who enlarged upon the idea of multicolor shadows on alien worlds in books such as *The Conquest of Space* (1949).

In honor of Flammarion's early lyrical descriptions of the visual possibilities of double stars' light, we suggest that the multicoloration of shadows on planets with different-color suns be called "the Flammarion effect." In many planetariums, you can see effects of this sort in displays that let you illuminate objects with several colored lights of variable intensity. You can even try for this effect at home with two colored light bulbs!

A schematic explanation of the Flammarion effect: two stars of different color cast different-colored shadows. The blue star casts a shadow to the right; only the light from the red star can fall into this area, giving this shadow a red tint. The reverse process colors the second shadow. The darkest area of shadow is where the two individual shadows overlap (center); no light reaches this region.

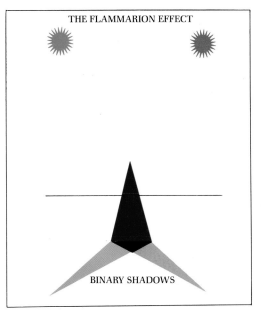

THE FLAMMARION EFFECT

BINARY SHADOWS

ALIEN PLANETS AMONG THE STAR SYSTEMS?

Among the 200 billion stars of our galaxy, there is only one star close enough for us to detect its planets. This is the sun, and we live on one of the planets circling it. A Jupiter-size planet circling the nearest star—one of the members of the triple-star system of Alpha Centauri—would be too faint to detect by present-day conventional telescopic techniques.

So the universe confronts us with a blatant question, blatantly unanswered: Is our planetary system a rare (or even unique!) product of unusual processes, or are planetary systems relatively common? The question touches both physics and philosophy. Its overtones are enormous. Did the solar system form by a peculiar accident that might befall only one star in a billion? Was it, as suggested by many cultures' ancient traditions, the special and unique creation of a God who planted only one Garden of Eden? Or is there an abundant variety of Edens, from watery worlds to methane swamps to airless dust-deserts? If there are other planets, how common are they? Are they habitable, either by ourselves or by somebody else? Could

creatures evolve to live in methane swamps? How unique is Earth?

These questions bring us face to face with a profound issue that we might describe as Copernican in scope. One of the key advances in human thought was the Copernican revolution of the 1500s and 1600s. Until that time, Earth was considered the center of the universe, explicitly as well as implicitly in people's minds. Ptolemy and the most influential great astronomers of antiquity taught that the sun, moon, planets and stars revolved around Earth. Copernicus proposed, instead, that Earth circles around the sun. The idea was controversial. Galileo helped it along by discovering with his telescope that moons orbited Jupiter. Here was direct observational proof that a *non*terrestrial body was at the center of some system. But when Galileo preached the Copernican model, the Roman Catholic Church halted his teaching and held him under house arrest for the last years of his life. The Copernican model was nevertheless vindicated and the Church's opposition to the new idea became a notorious scandal of humanity's intellectual progress.

Earth had now been displaced from the center of the solar system, but this was not the end of the story. Three related revolutions followed in succeeding centuries. They all made further displacements of Earth and humanity from the center of things. Even though Copernicus had removed Earth from the solar system's center, most people still assumed that *humanity* was the central, unique creation of the biological universe. Mankind was fundamentally unrelated to the "lower animals" and, indeed, had been appointed by God to be the baronial caretakers of creation. This smug world-view was severely altered, if not shattered, by the second revolution of displacement: the Darwinian movement of the late 1800s.

A double rainbow on an icy planet as we look through a cold mist of water droplets in a direction 180° from a double star. Each of the stars, which are behind us, produces its own rainbow; the brighter star produces such a bright rainbow (on the left) that we can see part of the outer arc with its reversed colors, as well as the conventional rainbow.

Ron Miller

A *strange eclipse. When our moon enters eclipse, it moves into the circular shadow of the Earth cast by the sun. But the planet's sun is a double star (which is out of the picture, behind us). The spherical planet, in the foreground, thus casts two circular shadows that overlap. At the distance of this moon, the shadow has a dark core where the two circles overlap; the dark core is surrounded by the remaining shadows, each illuminated by the light of only one star. If the stars were of different colors, the two shadows would display those two colors.*

Darwin's theory of evolution, supported by fossil after fossil found in subsequent decades, was that neither humans nor other species had been created full-blown. For most of Earth's history, there were no humans at all; other species "ruled the world." The whole universe of species had slowly evolved in ever changing patterns, with some species dying and new ones emerging as external conditions changed. Humans did not oversee the process from a lordly balcony like Marie Antoinette eating her cake; they were down in the meadows with the rest of the animals and plants. Species are interlinked, and if you seriously alter the environment or exterminate a few species, you're likely to affect the rest of the network.

The Darwinian revolution is not complete; rear-guard actions are being fought against it by anti-intellectuals. In the 1930s, for example, a few Soviet biologists, notably a researcher named Lysenko, tried to argue that evolution proceeded not just through more or less random fluctuations in the gene pool, followed by natural selection, but also through acquisition of traits during the organisms'

lifetimes. For example, if you studied hard to learn to dance, your daughters might *inherit* traits to make them good ballerinas. Soviet authorities adopted this as the party line and repressed free scientific testing of the idea. The motivation was ideological: the race (and perhaps Soviet society) could be improved by the chosen experiences of the New Soviet Man. Stalin loved Lysenko's theory. Today Lysenko's theory of evolution by acquired traits has been virtually abandoned.

Similarly, throughout this century, religious fundamentalists in the United States and elsewhere have maintained that Earth and humanity are a unique special creation and that evolution was not involved. Like the Soviets pushing the Lysenko theory, fundamentalists have tried to impose this theory on free scientific debate. Playing the same role as the Church's cardinals in the 1600s or the Commissars of the 1930s, they have attempted to ban the teaching of evolution in schools or to require that it be presented only along with an alternative "creationist" view. That view just happens to mirror their particular religious teaching, rather than, say, a selection of teachings from the creation traditions of Buddhism, Islam, Hinduism and other religions. Despite such attacks, for most of us in the later twentieth century the Darwinian revolution has once again displaced humans from a center that we occupied in Victorian times. We are now part of a larger whole.

A third and then a fourth revolution occurred around 1920. Until that time it seemed possible to believe that, even if Earth is not centered in the solar system, we might be centered in the Milky Way galaxy, and the Milky Way galaxy might be a unique system—a sort of galactic capital of the universe. In 1918, however, research by Harvard astronomer Harlow Shapley revealed that the solar system is off center in the disk-shaped Milky Way galaxy. Our sun was shown to be just one of many stars, not even near the center of its galactic system. During the 1920s, California astronomer Edwin Hubble obtained high-resolution photos of other galaxies (many of which had been erroneously called "spiral nebulae") showing that they were not nebulae of gas within the Milky Way system but rather separate systems of stars. They were far outside the Milky Way and comparable in size to our own galaxy. The Shapley and Hubble revolutions, as we might call them, were two more major displacements. The sun was just an ordinary star among billions in our galaxy, and even our stupendous Milky Way was not a unique center but just one of thousands of galaxies scattered throughout vast space.

The question of planets brings up the possibility, then, of a fifth intellectual revolution. Will it turn out, after all the displacements, that we cannot find any other planetary systems and have to conclude that we are unique after all? Or will exploration reveal a million planetary worlds, scattered in infinite variety across the sky?

SEARCHING FOR ANOTHER EARTH

The discovery of even one verified Jupiter-size planet beyond the solar system would be exciting. The discovery of a verified Earth-size planet would be a revelation. But astronomers want more: they seek to discover whole *systems* of planetary worlds, in order to measure whether the *distribution* of orbits and other properties is similar to that of the solar system. Such a measurement would allow us to assess whether a newly discovered system is indeed similar to the solar system, not only in the sizes of its worlds but in its actual mode of origin. Moreover, the discovery of a dozen such systems would begin to establish a statistical base, so that we could assess how common planetary systems are and whether their formation is related to the formation of multiple-star systems.

This brings us to the question: what is a planet? This deceptively simple question has been a controversial thorn in the side of the astronomical community. Because the discovery of a confirmed extrasolar planet would attract much interest, even those who try hardest to be objective and analytical want to be the first to announce such a discovery. Recall that below a certain critical mass, a newly formed starlike object would not achieve the central temperatures and pressures necessary to initiate nuclear reactions and thus could not become a true star—an object whose luminosity derives from nuclear reactions. As noted earlier, the critical mass is about 0.085 solar masses, or roughly 85 Jupiter masses.* There is a temptation to define any object below such a limit as a planet.

But as we saw at the beginning of this section, objects with a few tens of Jupiter masses would hardly look like planets. They would be red-hot from the heat of their contraction. They would look more like cool stars. Thus, as we saw, they are called substellar objects or brown dwarfs. So the real question is: where should we define the dividing line between a brown dwarf and a planet?

Suppose we start at the other end of the problem, with a body of one Jupiter mass, and imagine adding mass, working our way upward to see if we arrive at a point where we would no longer like to use the word "planet." Physical studies show that there is a transformation in the structure of bodies as we add mass. Imagine adding hydrogen to a Jupiter-like body composed mostly of hydrogen and hydrogen compounds. At roughly two Jupiter masses, the internal structure is crushed by the weight of the overlying material. As we add mass, the crushing causes

---

*Substellar objects and planets have such a small fraction of a solar mass that we will continue to discuss them in terms of Jupiter masses. Because Jupiter's mass is about one-thousandth of the sun's, Jupiter masses can be converted into solar masses by dividing by 1,000. Ten Jupiter masses, for example, equals 0.01 solar mass.

the body actually to decrease somewhat in size; if we continue adding mass, this tendency toward size decrease is overcome, and after perhaps five or ten Jupiter masses the size again begins to increase.

For these reasons, I propose that we arbitrarily define a planet as a body of less than two Jupiter masses circling a star. (The same terminology has also been adopted by a National Academy of Sciences committee studying the possibility of extrasolar planets.)

We might expect to find most planets in circular orbits and lying in the plane of the parent star's equator, because the sun's planets have virtually circular orbits close to the plane of the sun's equator. But this is certainly a prediction that needs to be tested by observations. If we find a planet in a noncircular orbit, or in an orbit highly inclined to that of its star's equator, we will ask whether that particular planet formed in a markedly different way from the sun's planets.

In the few cases of known systems of co-orbiting low-mass stars or substellar objects where orbital statistics are available, the orbits are usually not circular and co-planar. This reinforces a suspicion that most known multiple-star systems and substellar objects really did form by a different process from our solar system.

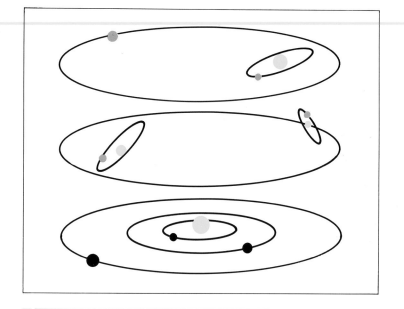

*T*ypical orbits in multiple-star systems place two stars in a tight orbit and an additional star or pair circling the first pair at great distance (top and middle). The orbital plane of a second pair is usually different from that of the first pair (middle), and the orbits may be elliptical. Such systems are hierarchical. In contrast, planets circle the sun in nearly circular orbits in the same plane (bottom), each space roughly twice as far from the sun as its inward neighbor. If planet-size bodies are discovered orbiting other stars, it will be important to examine their types of orbits to determine whether these bodies were formed in the same way as our planets in the solar system.

HOW PLANETS FORM

I f we hope to discover planets around another star, we should begin by asking how planets came about in our own solar system—the only system in all of space where we *know* that planets exist. Planetary scientists, sifting clues contained in planetary orbits as well as in terrestrial rocks, lunar rocks, the fragments of asteroids we call meteorites, and even in two or three "meteorites" recently identified as probably blown off Mars, are archaeologists of the solar system. They try to reconstruct its early days.

The various rock samples yield dates proving that all the planets and interplanetary bodies formed at about the same time, 4.5 billion (4,500,000,000) years ago. At this time the sun had just formed, having completed its col-

William K. Hartmann

Inside a disk-shaped nebula of dusty debris orbiting around a newly formed star not seen. The light of the star, from our vantage point in the nebula, is partially obscured and reddened by the nebula's dust. From a larger distance the star may be totally hidden by dust clouds, but the dust swarm might be revealed by its infrared radiation as it re-radiates the light it has absorbed from the central star. The dust grains may aggregate to form small planetesimals (in foreground) that may accumulate into planets in a few million years or less. Chemical and geologic data from meteorites indicate that our solar system formed in this way. This scene from our solar system's past may be repeated today at thousands of sites in our galaxy, in cocoon nebulae around known young stars.

nside the dusty cocoon nebula near a young star, grains of dust have been colliding and aggregating into asteroid-like worldlets called planetesimals. This same process happened in our own solar system some 4.5 billion years ago. We stand here on the cratered surface of one such object. As the dust clears due to aggregation into planetesimals, the nebula becomes more transparent than it was initially, allowing a view of the coronal streamers around the very active young star shown here eclipsed (or, more probably, occulted) by a passing planetesimal.

lapse as a protostar and just started its hydrogen-burning life as an ordinary single star. The interval of the planets' formation was comparatively short, only some 10 to 100 million years. The data show that about 4.5 billion years ago the sun was surrounded by a cloud of leftover debris—gas and dust. By about 4.4 billion years ago some of the gas and dust had aggregated into planets, while the rest had dissipated. How did this happen? Do the clues suggest that this could have happened around other normal stars, or did it require unusual circumstances?

The data—a combination of theoretical models of protostar collapse plus chemical and age measurements from the oldest rocks—indicate that the process was probably something that could have happened around many stars, or at least around many single stars. Dust grains were common in the dust surrounding the primordial sun. Observations of stars support this; we can see such grains in nebulae around other young stars. They are called *cocoon nebulae* because they resemble the caterpillar's

shroud, cast off as the butterfly emerges. Similarly, the cocoon nebula hides a protostar in a wrapping of opaque gas and dust. As the star "turns on," however, the cocoon of dust is blown away. The cocoon nebulae discovered around distant protostars appear to duplicate our own solar system's original nebula, so that the sky offers us natural laboratories where planet-forming processes may still be going on today.

If only we could see these processes in enough detail to be sure what is happening around these stars! We believe that both in our own system 4.5 billion years ago and in the remote cocoon nebulae, dust grains condensed as the gas cooled. This happened much in the way that ice crystals condense in high-altitude air as it cools during ascent, perhaps over a mountain. As they circulated around the sun, the dust grains collided at low speed. If they bounced apart, there was no net effect, but sometimes they remained in contact—perhaps initially due to electrostatic forces, the same forces that make dust cling to phonograph records. As soon as loose conglomerates of a few grains formed, these were efficient in capturing adjacent grains during subsequent low-speed collisions. A similar process can be seen during a heavy snowstorm: many of the falling snowflakes are conglomerates of several flakes that collided and stuck together. The early nebula around the sun was like a blizzard of dust grains! Evidence for this stage of events comes from both lab experiments and study of the microscopic structure of meteorite materials.

A second growth mechanism has been suggested. In some regions of the primordial solar nebula, there were enough grains close enough together that their mutual gravitational attraction caused individual clouds of grains to collapse into weakly bonded "flying dustballs" in the same way that gas clouds collapse into stars. Calculations suggest that such dustballs in our own solar system could have reached diameters of a few kilometers by gravitational collapse of dust clouds alone, even without the slower, grain-by-grain accretion process.

The little asteroid-like bodies that grew by such processes are called *planetesimals*. A planetesimal that reached even the size of a tennis ball would have been quite efficient in sweeping up adjacent grains because it would have had a loose, granular structure that absorbs energy on impact, stopping the motion of an incoming impactor and preventing rebound. Thus a particle that collided with a larger, granular planetesimal would probably have been captured into the rubble on the larger body's surface. To demonstrate this effect for yourself, go out and drop a small rock on a concrete sidewalk. It will bounce. Now sprinkle some fine dust or kitchen flour on the sidewalk, to a depth of only about half the diameter of

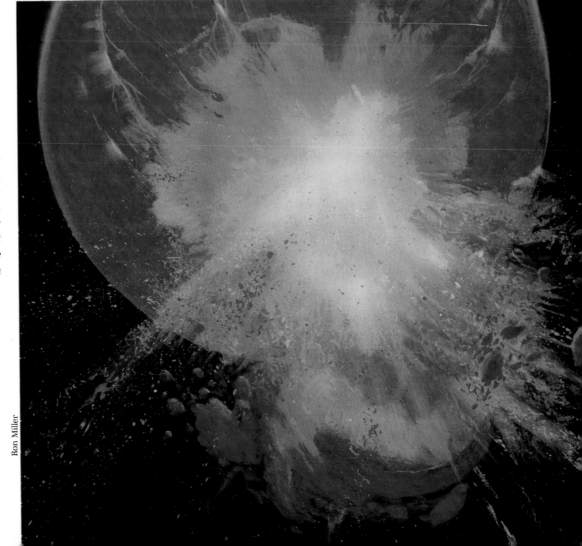

ollision! We know from studies of the moon and meteorites that as planets formed in our own planetary system, the surfaces and interiors of many of them were heated to molten conditions. We know also that pre-planetary bodies of various sizes have collided—the process that allowed asteroid-size objects to accumulate into planet-size objects. The same process may be occurring today in the dust-rich nebulae detected around many newly formed stars, where a collision between two partially molten bodies would be a spectacular event when viewed from nearby. Here, a large asteroid has just blasted out parts of the mantle and molten crust of a growing planet with a thin atmosphere. While most of the debris will fall back, a fraction may remain in orbit to form a ring or a moon.

Ron Miller

Ron Miller

Killer comets. Comets are the icy debris of planet formation, ejected to the outskirts of planetary systems by gravitational encounters with the planets. It is believed that passing stars, or remote orbiting companion stars in binary systems, could deflect large numbers of comets into the inner parts of a planetary system, causing intense bombardments of the planets. Here, at sunset on such a planet, we see an evening sky filled with beautiful but unwelcome celestial visitors. They pose a potential threat to the planet's lifeforms because a comet's collision with a planet could alter climates and radically change the future history of species evolving on the planet. Any one of those comets may eventually collide with the planet on which we are standing.

the rock. Drop the rock on this granular surface, and you'll find that it hardly bounces at all. Such experiments, and more sophisticated laboratory data, show that even a little surface dust would help a planetesimal grow. Once a few large, granular planetesimals "pulled ahead" of their companions in the race to reach larger size, their gravity would help in sweeping up smaller material. The larger a planetesimal grows, therefore, the faster it will continue growing. A "runaway" growth effect begins.

Thus the early generation of kilometer-scale bodies, whether formed by slow, grain-by-grain accretion or by rapid gravitational collapse, would have continued to grow efficiently by the grain-by-grain accretion process. Eventually, a few large bodies had to emerge, as long as the process was not disturbed by any other effects.

On the basis of this reasoning, many researchers believe that a swarm of dust grains around a star naturally evolves into a swarm containing a few larger bodies and a multitude of smaller bodies. For example, at one stage there might be one body that has reached a diameter of 1,000 kilometers (about a third the size of the moon), a couple of bodies around the 500-kilometer size, a hundred bodies around the 100-kilometer size, and thousands of 10-kilometer bodies. This is no imaginary picture; this is exactly the situation in our own solar system's asteroid belt. Many researchers believe that the asteroid belt is a fossilized remnant of our own long-gone planetesimal population. In the belt, growth processes apparently were halted by the gravitational disturbances of nearby Jupiter. The existence of the asteroid belt supports the general theory of dust-grain accretion.

The terrestrial planets, such as Earth and Mars, probably grew to their present size by this process. The development of the giant planets involved another step. As they were growing, they accreted not only the rocky material of their surroundings, as did the terrestrial planets, but also additional icy material, which was more common in their cold, outer parts of the solar system than in our own, warmer region. Thus, whereas Earth's growth leveled off at one Earth mass, the giant planets' ancestral embryos grew to one Earth mass, then to two Earth masses, and then to ten Earth masses. At this point, something new happened. The gravity of the ten-Earth-mass core was so strong that it began to pull in and capture gases from the surrounding nebula. These gases were essentially the same materials, mostly hydrogen and helium, from which the sun itself had originally formed. Thus the giants leaped ahead to much more massive sizes. Uranus and Neptune grew to around 15 Earth masses, while Jupiter attained a whopping 318 Earth masses!

Although the largest planet in our system reached just one Jupiter mass, there is no reason that, in some particularly massive circumstellar nebula, these processes could not have produced a substellar object of perhaps twenty Jupiter masses—a brown dwarf larger than Jupiter but smaller than the sun, orbiting around the central star. In principle, the same processes might occasionally produce a small star of perhaps 100 Jupiter masses as a companion to a larger star.

Could there be, say, three-Jupiter-mass worlds of rock, ice and gas in some systems that formed in this way, while in other systems there are strangely different three-Jupiter-mass worlds of hydrogen and helium that formed directly by gravitational collapse? Such a question shows why it is of special interest, as we study the smaller companions to other stars, to discover not only their existence but also clues as to which of the many possible formation processes was involved. As mentioned, one such clue is whether the orbits of such bodies are circular and in the plane of the larger star's equator (as among the solar system's planets) or elliptical and inclined (as in several known star systems, which may have formed by direct collapse instead of collisional accretion).

The formation modes we have been discussing could have happened near many stars, meaning that planets may be fairly common. Seventy years ago, many astronomers had a different view. They thought that planets formed in an unlikely, accidental way, by aggregation of debris blown out of the sun by a rare, chance collision with a passing star. It's easy to calculate how often such collisions might occur, given the spacings and velocities of the stars. They turn out to be extremely rare, and if planets could not form without them, then planetary systems would be extremely rare in our galaxy. The modern theory, buttressed by meteorite studies, lab work and dynamical calculations, has just the opposite ramification. It suggests that systems of dust around many newborn stars easily aggregate into planets; planets ought to be common.

The surface of this infant, Earth-like planet has been heated by impacts of the planetesimals from which it formed. Now the last stages of pervasive eruptions are covering the planet's surface with overlapping lava flows—forming the first stable surface. Rising above the horizon, a moon that formed near the planet is slowly moving away from the planet due to gravitational forces between the two bodies—following the same history as Earth's moon.

CATASTROPHES AMONG PLANETS

The reasoning above gives the basic modern picture of planet formation. It makes the growth of planets seem relatively straightforward, but there are, as in any theory, some subtleties and complications. One involves the collision velocities. Each zone of the new planet-forming system contains not only one large "runaway" embryo but also a second-largest planetesimal (perhaps half the size of the first), a third-largest, and so on. Each of the smaller planetesimals may eventually collide with the largest, in a spectacular, explosive collision. As long as the incoming body does not hit its target at too high a speed, the collision does not blow the target planet to smithereens. It does blow it, one might say, halfway to smithereens: the target planet may be totally fragmented into a cloud of small pieces. But there is not enough energy to accelerate these pieces fast enough to escape each other's gravity; they simply fall back together again. All the landforms, rocks, continents, even oceans of the target planet are gone forever, yet a new planet with its own new craters, boulders and canyons re-forms in its place within a few millennia!

Suppose life had started to evolve on such a planet. It is haunting to imagine whole hierarchies of lifeforms on the planet vanishing from the universe in a climactic hour, and quite different ones evolving on the "same," renewed planet a hundred million years later. Perhaps this has happened in other systems, but in our solar system we believe the major collisions all happened in the first few tens of millions of years before substantive lifeforms evolved.

There is another aspect to the velocity problem. A planetesimal is much more likely to make a near miss to a target planet than to make a direct hit. (As dart players know, the area of the rings around the bull's-eye on a dartboard is much greater than the area of the bull's-eye itself.) As the largest planetesimals in a swarm grow, the smaller ones will probably have near-misses with large ones before they collide with one of them. They will be deflected onto new paths by the targets' gravity fields. As a result, the smaller ones gradually have their orbits disturbed and acquire higher velocities relative to their neighbors'. The average velocity in the swarm thus creeps upward and eventually reaches the point where some planetesimals can shatter substantially larger-size target planetesimals during collisions, blowing them apart so violently that they never reassemble. In effect, the swarm of smaller planetesimals starts grinding itself down into smaller fragments, even as the largest one is still growing toward planethood.

This is the status of our solar system's asteroid belt today. The largest asteroid, Ceres, is probably fairly safe from being destroyed, but the smaller asteroids are grinding each other down into dust. When a pea-size asteroid hits a mountain-size asteroid in the belt, it blows off the equivalent of several hundred peas, resulting in net mass loss. When a city-size asteroid hits a mountain-size asteroid, it may shatter the whole mountain.

Conceivably, there could be systems in which planet formation is prevented by this erosion. One giant planet might form first and disturb the orbits of the neighboring dust to the extent that all subsequent collisions lead to mass loss instead of mass gain. Or a passing star might disturb their motions in a similar way. A dust cloud, or cocoon nebula, might be seen around such a star, but it might grind itself away to nothing in a million years.

What happens to the fine dust sandblasted off the planetesimals? All young stars apparently have strong "stellar winds": outrushing gas shed off the central star, analogous to the "solar wind" that sweeps out from the sun through our own solar system, blowing comet tails away from the sun like windsocks. Once it gets small enough,

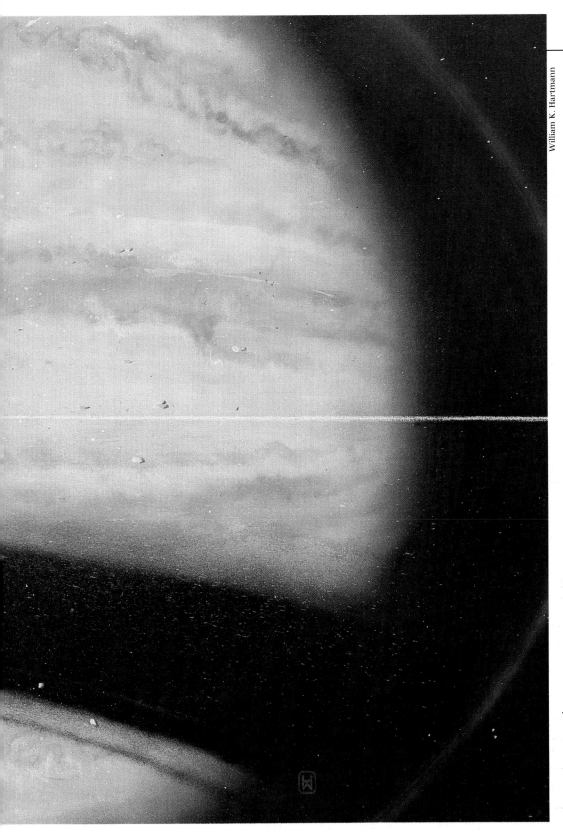

William K. Hartmann

One catastrophe that can befall worlds is a major collision. Voyager team scientists concluded that some of Saturn's inner moons have been blasted apart and reassembled several times by large impacts. Such an event, shown here near a ringed, Saturn-like planet, may be occurring among alien planets even today. Large asteroids or comet-like bodies, formed at the same time that the planets grew, wander among young planets and have one of two fates: either they make a near miss to a planet and are thrown out of the system by gravitational forces; or they actually hit a moon or the planet itself, causing a cataclysmic disruptive event. Debris ejected from the collision may be added to the ring system and reaccumulate to form a new satellite, or it may fall onto the planet.

Pamela Lee

A condemned planet. A red giant star has expanded until its thin, outer atmosphere is about to engulf a planet in an orbit like that of Earth. The star surface is mottled with cooler and warmer regions, all radiating with temperatures around 3,000° F. The star covers much of the planet's sky with a blazing inferno, causing the liquids on the planet to boil away and the atmosphere to stream off into space like a comet's tail. Skimming through the star's outer atmosphere, the now sterile planet will experience drag forces and ultimately spiral into the inferno of the star's interior.

the dust will be carried along by the solar or stellar wind to be dispersed into interstellar space. In this way, the eroded dust would be blown out of the solar system as fast as it formed. Thus a planetesimal swarm that once had a shot at making planets might, if its velocities got too high, erode and have its debris swept away forever, leaving the star planetless. Or a ring of asteroids and pebbles may stay behind in orbit around the star.

Aside from direct collisions, there are many other possibilities for planetary disasters. A planet that undergoes a close approach to a large passing body may escape a collision but still suffer consequences. As mentioned above, the mutual gravitational attractions of the bodies alter their orbits. After the near-miss, the orbit may be more elliptical, passing closer to the star than it did before or farther away from it. The temperature regimes on the planet would be drastically changed in the course of only a few months or years. The rotation rate of the planet might change as well. If the orbit change is drastic enough, the planet might be kicked out of its star's system and wander in interstellar space as a derelict.

Such changes would be most likely to occur early in the planet's history, when other planetesimals are still around. But the potential for disaster lurks in the planet's late history, too. As the parent star evolves, its energy output may fluctuate, or steadily increase, or decline. The planet of a star turning into a red giant or a supernova is, obviously, doomed to serious alterations.

Such drastic changes are uncommon. If we could tour the galaxy in an imaginary hyperspace cruiser and discover 1,000 planetary systems, the chances are that most of them would not experience catastrophic changes in the next thousand years or so. On the time scale of human history, planets seem stable. We think of our planet's surface as a static stage on which our human dramas unfold. But on the time scale of biological evolution, or of astronomical evolution, planets are extremely variable. On Earth alone we have seen ice ages on a time scale of 20,000 years and catastrophic climate alteration induced by asteroid impact on a time scale of 100 million years. We expect the sun to change drastically—to swell up into a red giant—in the next six billion years or so. "All the world's a stage," said Shakespeare, but neither his Globe Theatre, nor Earth itself, stayed unchanging forever.

William K. Hartmann

A planetary disaster... or a new start? As a planet forms, it is likely to be only the largest of a number of planetesimals, each growing by accumulating smaller bits of debris. During its final stages of growth, the planet may collide with neighboring planetesimals that have also grown to large size. Here, an Earth-like planet is being struck by a moon-size body. The drama of collision and ejection of debris will take an hour or so to unfold. Although the planet itself will survive, primitive lifeforms that might have developed on the planet will be demolished. Much of the planet's crust will be disrupted, and some debris from the planet's and the planetesimal's outer layers will be blown into a ring system around the planet. Here, the planet's shadow is cast through the expanding debris-cloud. Many planetary scientists theorize that Earth's moon originated by such a process, forming in Earth-orbit from dust blown out of the Earth. What seems a disaster for the original planet may be a birthing process for a new world. [Left]

A lost planet and its moon. Like Pluto and its large moon Charon, this double planet originally circled its massive parent star at a large distance. The massive star rapidly burned its fuel and blew off much of its mass in a supernova explosion. Its resulting weaker gravity was inadequate to hold the planet, which now found itself moving faster than escape velocity for its particular orbital position. As a result, the planet and moon, still circling each other, sailed off into the interstellar star cloud that originally spawned the massive star. The pair is lit here by an offstage orange giant that is passing at some distance out of the picture to the right. [Right]

DO ONLY SINGLE STARS HAVE PLANETS?

You will recall that at least half the stars, or star systems, are double or multiple systems in which more than one star has formed. If you consider the collisional motions of the dust grains trying to aggregate into a planet, and then picture a second star suddenly being put in the midst of the system, you can imagine that the second star's strong gravity will seriously disturb the motions of the dust (and of any planet that might be forming). One effect would be to accelerate the motions of some grains, making them collide faster with other grains so that it's harder for them to stick together. In fact, the collisions could then be *destructive* instead of *constructive*; the faster collisions would tend to blow planetesimals apart, as discussed above. A second effect is that some planetesimals (or even full-grown planets!) could be ejected from the system altogether into interstellar space, much in the way that one of the Voyager space vehicles was ejected from the solar system after passing close to Saturn.

The net effect is that planet growth is somewhat restricted in double- and multiple-star systems. Planets in double-star systems could grow in a zone close to either or both of the two stars, where the gravity of only one star dominates; they would probably not grow halfway in between, where the orbits would be disturbed by gravitational pulls from both stars. Alternatively, if the two stars are close together, planets could form at a large distance from both of them, but not nearby. In other words, if Jupiter were a star, "planets" such as Mercury or the moons of Jupiter might exist, but Mars or Saturn might be unstable. The complex gravitational forces of the double-star system might prevent dust particles from aggregating

in orbits corresponding to the position of these planets; or, if aggregation did proceed, the resulting planet might ultimately be kicked out of the system by the same forces. Similarly, if Mercury were a star, Neptune might have formed but Venus might be unstable. For these reasons, planetary systems seem somewhat more probable around single stars than in double- or multiple-star systems.

Nonetheless, it's plausible that planet systems might grow close to one star in a widely separated pair of stars. A curious, remote star system, more than 4,000 light-years away, may clarify this situation. It is a double star called Epsilon Auriga. Every twenty-seven years, it fades into eclipse for nearly two years! The eclipses of Epsilon Auriga, which happen only once or twice during each astronomer's career, have been studied carefully. The last

The unusual double-star system of Epsilon Aurigae. The huge, hot, massive star in the background is as big as Mars' orbit. It is partially obscured by a disk of dust surrounding a hidden companion star of unknown nature, probably a very massive, still forming, young star of several solar masses. The hidden star is as far from the visible star as Neptune is from our sun. It is apparently surrounded by a disk of heated dust, which passes in front of the visible star every twenty-seven years, partially obscuring it from Earth. Some astronomers suggest that planetary bodies might be forming in the dusty disk. It is as if we had a second planetary system forming around Neptune in our own system.

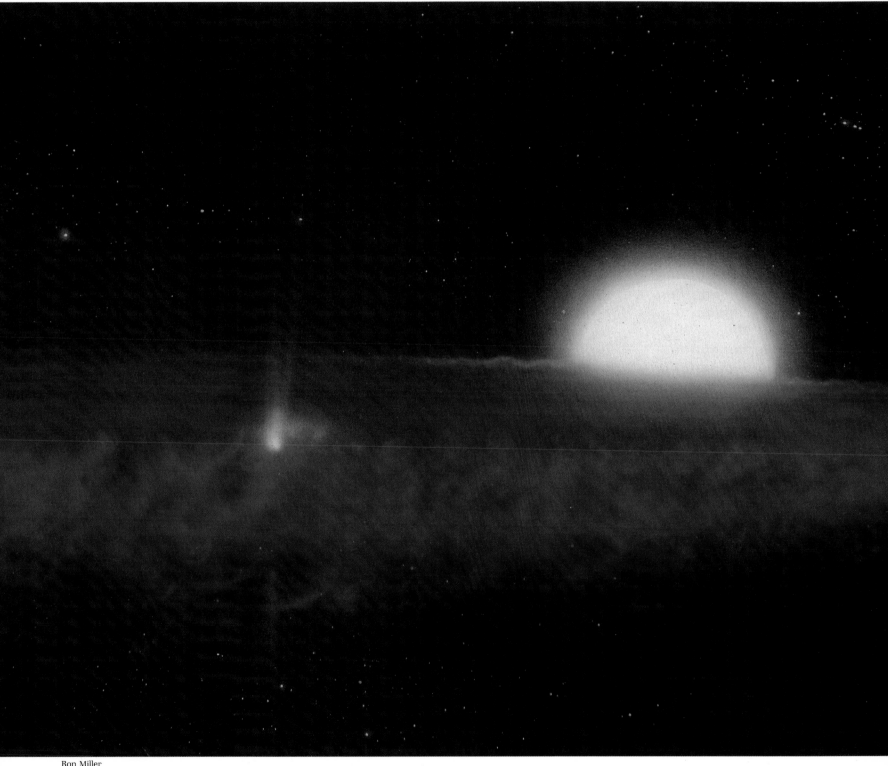

Ron Miller

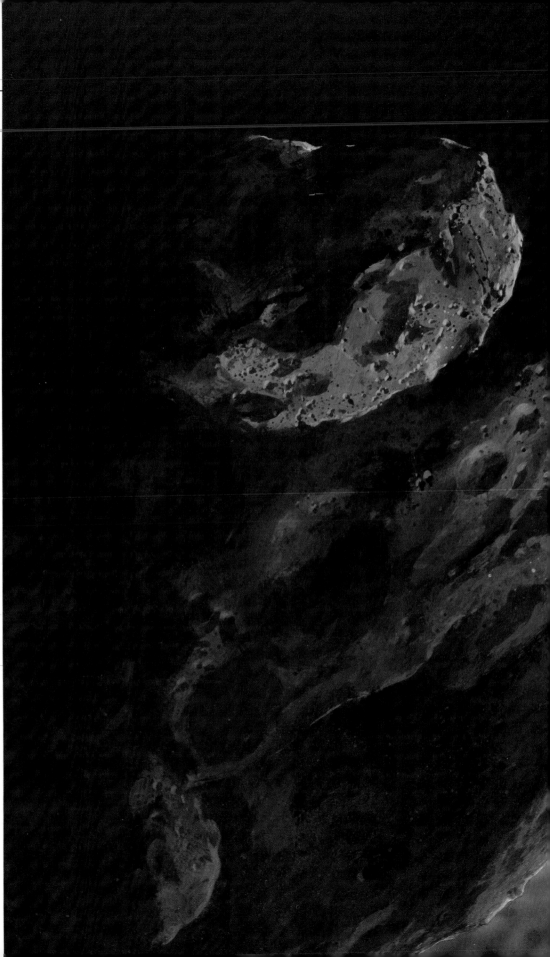

Pamela Lee

One theory about the unusual star Epsilon Aurigae is that planetary bodies may be forming within the disk-shaped nebula orbiting around it. As it orbits about the prominent supergiant, this cloud occasionally passes in front of the star, and the thick load of dust and rocky planetesimals helps diminish the star's light, causing two-year-long eclipses that have been monitored from Earth. The possible presence of a planet-growing process in the nebula makes the system interesting for future studies despite its great distance, some 4,000 light-years, from Earth. Will the cloud collapse, leave a residue of planets or simply disperse?

one was in 1982–1984. One star is a hot, bright, massive star, but the object that passes in front of it and causes the eclipses is invisible. Apparently, the companion star is even more massive, totaling as much as twenty-eight solar masses, but is hidden in a thick disk of dust. As the two objects orbit around each other, the dust disk passes in front of the visible star every twenty-seven years, blocking much of its light. The disk is centered about as far from the visible star as Neptune is from the sun, and the disk has dimensions about half the size of our solar system. Therefore, astronomers have suggested that the dust in the disk might be forming asteroids or planets. Perhaps by the time of the next eclipse, in 2010, we will have giant orbiting space telescopes that can tell us in more detail what is happening in this intriguing double-star system.

The natural tendency of dust around young stars probably is to aggregate into planets, but we are not sure. While it seems almost difficult to imagine circumstellar dust clouds *not* forming planets, we raise the caveats of velocity and multiple-star disturbances as examples of the idea that stars do not *necessarily* spawn planets. Is our planetary system, then, one of many, or is it just a lonely, cosmic curiosity? We need to search the night skies with our instruments to find out if other stars have planets!

FINDING EXTRASOLAR PLANETARY SYSTEMS

The race is already on to find planets near other stars. Several techniques are being developed to allow detection of tiny, almost invisible masses in orbit near stars. Some of them are offshoots of other astronomical work dealing with stellar properties and are already in operation.

One technique that might at first seem promising is nearly hopeless. This would be to try to get direct photographs of a planet in orbit near another star. The problem is that the glare of the star overwhelms the extremely faint planet. This technique may eventually be used on nearby stars by applying new imaging technology and giant telescopes orbiting in space, where lack of atmospheric shimmering facilitates very clear images, but its effectiveness is much reduced for very distant stars. Instead of this direct approach, we can list four more subtle but powerful techniques that will be used in the next decade or so.

First is the method called *astrometry*, which is the study of the motions of stars. If a planet (or a star or any other object) orbits a star, then the two bodies are really orbiting around the center of gravity of the system. Just as a stellar companion of an unseen black hole may perform orbital motions that reveal the presence of the invisible black hole, a star that parents an unseen planet is itself describing a relatively small orbital path as its companion, the planet, moves around it at a greater distance. The more massive the planet, the greater the star's oscillation and the greater the chance of detecting the oscillation by extremely precise position measurements. The oscillation reflects the mass and distance of the orbiting planet even though the planet itself cannot be seen. Already, astrometry has been used to detect unseen stars and substellar objects around many nearby stars.

Among the handful of closest stars, the technique is powerful enough to detect objects of a few Jupiter masses, or perhaps even less, in the planetary mass range. It has successfuly confirmed several low-mass stars, of masses around 100 or 200 Jupiter masses, orbiting around some nearby stars. In the 1970s, Dutch-American astronomer Peter van de Kamp thought he had detected one or more planets, perhaps as small as 0.6 Jupiter masses, around the third closest star, a system called Barnard's Star. But later observations, just as sensitive, failed to confirm the planets and most astronomers now doubt their existence. The method is being improved, however, and it should be possible within a decade to detect planet-size objects orbiting a number of nearby stars . . . if they're there.

A second technique for discovering extrasolar planets is to seek eclipses. Astronomers carefully monitor the brightness of selected stars from day to day and watch for

Ron Miller

A possible but controversial brown dwarf. This view is based on reports about a faint, red star (far background) called VB 8, only a tenth as massive as the sun and twenty-one light-years away from the sun. Naval Observatory astronomers in 1983 reported evidence from this star's motions that it has a companion, its mottled surface filling the foreground. Arizona astronomers subsequently announced that on several occasions they got direct observational indications of this "small" companion object, only several times more massive than Jupiter. Too small to be a star and too big to be a planet, it was thought at first to be the best-known example of a "substellar object," or "brown dwarf." The temperature of this object was estimated at 1,400 K (2,061° F), and it would perhaps glow with a dull light emanating from below a thick atmosphere in which some materials might condense to form cooler, darker clouds. The excitement of this evidence for a "missing link" between planets and stars dimmed more recently when other teams of astronomers were unable to confirm the detection. The VB 8 system is presently a mystery awaiting newer, better observations. Does the brown dwarf really exist?

William K. Hartmann

N ot all Earth-like worlds need be planets in their own right. Saturn's system testifies that large moons with atmospheres can circle giant planets. Here we visualize an Earth-size desert world circling a ringed planet, which is faintly visible beyond the clouds. Methane-produced organic condensates in the lower atmosphere produce peach-colored cumulus clouds, while high, white cirrus clouds of water ice crystals float high above.

a dimming that would signify that a small, dark object had crossed in front of it. The trouble with this technique is that enormous amounts of observing time are required. Even if many other stars had "solar systems," the chance of finding one where the planets' orbits were seen edge on, passing across the disk of the star, is small. However, this technique might be brought in as a backup if the other techniques revealed that a certain star had unseen companions in nearly edge-on orbits. Detection of an eclipse would give good information about the size of the unseen body, since the amount of light blocked by the planet would depend on its area.

A third technique of planetary detection is to seek not star motions associated with the presence of a planet, or the too-faint-to-detect reflected light bounced off it from the star, but rather the *infrared heat radiation* from the planet. There is a cunning application of physical principles involved here. Because hotter objects emit bluer light and cooler objects emit redder light, objects as cool as a planet emit no visible light but do emit long-wavelength light, called *infrared light*, which is too red to be seen by the eye. A star emitting mostly visible light gives off very little of this infrared light. Thus an astronomer seeking a planet uses a telescope equipped with an infrared detector (not unlike that on heat-seeking missiles) that is insensitive to the visible starlight. The detector therefore hardly "sees" the glare of the star at all and has a chance of picking up faint infrared emission from a planet or a very low-mass star orbiting around the primary star.

One apparent success of this technique came in 1984, when Arizona astronomers used infrared imaging techniques to observe for the first time a substellar companion that had earlier been suggested by the astrometry technique in 1983. The astrometry had indicated that a companion to a star known as VB8 (star number 8 on a list studied by the Belgian-American astronomer Van Biesbroeck) had a tiny companion of uncertain mass. The infrared work indicated a mass perhaps as small as five Jupiter masses and a temperature of about 1,400 K (2,060° F). A reversal came in 1986, however, when other astronomers searched for VB8 B, as it had come to be called, and were unable to find it! Could it have moved partway around the star to a position where it was no longer visible? Calculations suggested that this was not likely in the time available. Astronomers vowed to continue the search around VB8 for the suspected but now officially unconfirmed substellar companion.

A fourth method involves a phenomenon called the *Doppler effect*. An example of the Doppler effect is the apparent lowering of pitch of a train whistle or car engine as it speeds past you. The wavelength of the sound you hear is a little higher as the object approaches and a little lower as it moves away. In the same way, the wavelength of light you see is a little higher, or bluer, when an object approaches, and a little redder as it moves away, although in the case of light the effect is so small that we don't detect it in everyday life. Astronomers, however, can detect and use it to measure how fast neighboring stars or galaxies may be moving toward or away from us.

A star that has an unseen planet moving around it "wobbles" not only from "left" to "right" in the sky but also "back" and "forth," toward and away from us, along our line of sight, as it is tugged by the gravity of the orbiting planet. Using the Doppler effect, astronomers measure the wavelengths of light from the star and determine whether it exhibits any orbital motion back and forth along the line of sight. A beauty of this technique is that it works just as effectively at any distance, as long as the star is bright enough for astronomers to make good readings of Doppler shifts. The ultimate limitation on this technique is that

motions of the light-emitting gas in the surface layers of the star cause their own Doppler shifts, which may mask the subtle orbital motions caused by a small planetary companion.

This fourth method has led to some exciting results. In 1987 Canadian astronomers announced that after monitoring sixteen stars in this way, they detected possible unseen companions of only one to ten Jupiter masses around seven of the stars. Two of these detections were especially strong and provocative. The nearby star Epsilon Eridani, only slightly less massive than the sun, has a companion of only about two to five Jupiter masses. A more distant star, Gamma Cephei, seems to have a tiny companion of about 1.7 Jupiter masses; it could be a true planet! Astronomers will rush to make further studies of those and similar targets!

The next decade will see radical advances in sensitivity, especially if these techniques can be tried with orbiting telescopes in space or even with ground-based telescopes that are dedicated solely to the program. My own guess is that by 1995 we will have confirmed Jupiter-size or smaller worlds orbiting one or more stars. If so, then we will begin to be confident that planetary systems have evolved around many stars, and we will have an even stronger suspicion that we are not alone in the universe.

On the other hand, if I am wrong, we may find that the number of substellar objects dramatically cuts off as we get into the planetary mass range. If we search 100 systems with a sensitivity adequate to detect Jupiters or Neptunes, and find nothing, we will have exciting and eerie new questions: Are we alone after all? Is something wrong with our theoretical understanding of planet-system formation? The search for planets has to have provocative results, no matter how it comes out!

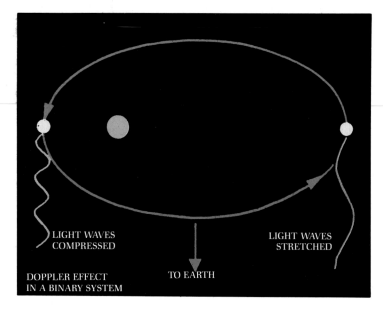

LIGHT WAVES
COMPRESSED

LIGHT WAVES
STRETCHED

DOPPLER EFFECT
IN A BINARY SYSTEM

TO EARTH

A s one star orbits around another, its light waves undergo the Doppler effect. This is a change of wavelength, depending on whether the star is moving toward or away from the observer. If the star moves toward us (left), more waves per second are perceived by us, giving an apparently shorter wavelength than normal. Similarly, if the star moves away (right), the wavelengths are perceived as longer.

Ron Miller

O n a hypothetical planet of a brown dwarf. Above the crags at the left floats a substellar object: too small to be a true star, but too big to be a true planet. It is red-hot, glowing from the heat generated by its own contraction, but it gives only faint light and slight heat to the planet. Most of the planet's light comes from the major body in this imaginary system, a true, sunlike star, located about twenty times as far from the planet as the Earth is from the sun. If such a system were discovered, astronomers would be hard-pressed whether to classify the star and the brown dwarf as essentially a double-star system or to regard the brown dwarf and planet more as a large planet and satellite in a single-star system. The question may be more than semantic; there may be different physical formation processes for stars and planets. Indeed, some recent evidence suggests that brown dwarfs are very rare; they may fall in a mass range that is hard to find.

Satellite and ground-based observations have revealed that the seemingly normal star Beta Pictoris is surrounded by a hitherto unknown disk of dust. Properties of the disk suggest that planets or asteroids could have formed in the inner parts of the dust disk, closest to the star. The dust system, which has been indistinctly photographed from the Earth, extends about 400 astronomical units from the star—a distance matching that at which many comets orbit our own sun, far beyond Pluto. Beta Pictoris is about 50 light-years from our solar system.

NEW DISCOVERIES

In 1983, the Infrared Astronomical Satellite (IRAS) began looking at various stars and discovered that a few ordinary, evolved, hydrogen-burning stars are surrounded by swarms of dust particles. These were not newly formed stars of the type long known to have dusty cocoon nebulae. They were evolved stars, like the sun, that were supposed to be solitary. Vega is a notable example. This well-known star, prominent in the summer sky and about three times as massive as the sun, is thought to be a billion or so years old—plenty of time to disperse its primordial dust cloud—but has a swarm of 90 K (300° F), millimeter-scale particles around it. The total mass of dust is uncertain, but it's at least as much mass as the moon. Another star picked out by IRAS as having infrared-emitting dust was the brightest star in the Big Dipper. What could the dust mean? Where was it coming from? Was it part of a system of planets, just as cometary and asteroidal dust is a by-product of our own planetary system?

A star with an even thicker dust-disk was Beta Pictoris, the second brightest star in the southern constellation of Pictor, and detailed study of it helped answer some of the questions. In January 1984 Arizona astronomer Bradford Smith, who is widely known for leading the Voyager Imaging Team's study of the outer planets, and his California colleague Richard Terrile, used advanced equipment to photograph Beta Pictoris' dusty disk. The disk lies edge-on to us and extends 400 astronomical units from the star, or roughly 10 times the distance of Pluto's orbit from the sun. Further studies of the properties of the dust revealed that many of the particles were the size of pebbles, vastly larger than the microscopic dust grains that exist in interstellar space. This proved that at least some aggregation had occurred, although astronomers could not directly determine whether the particles in the system were still aggregating—growing toward planetary size, perhaps—or eroding. Conceivably, there could already be some asteroid-size bodies in the system, and the dust we see could be produced by their collisions with meteorites. The brightness distribution of the dust-disk and star suggested to astronomers that the inner tenth of the disk, a region about the size of our planetary system, has less dust. Therefore, some astronomers suggested that the dust in that region may have already aggregated into planets. Beta Pictoris is estimated to be a moderately young star, perhaps a few hundred million years old, and twice as massive as the sun. Perhaps it is in the final stages of forming a planet system and its excess dust still forms a broad disk stretching beyond its planets in the same way our cloud of comets stretches beyond the solar system.

The discovery that we could directly study dust systems possibly growing into planets has led to further aggressive work by astronomers in California, Hawaii, Wyoming, at Baltimore's Space Telescope Institute, and in Europe and elsewhere. By 1987, astronomers had discovered dust around a number of nearby middle-aged stars that had been thought to be single and had published new images of similar dust-disks around other stars known to be young. These included T Tauri, a long-known prototype of a young star system with a dust cocoon and a very small infrared companion, and HL Tauri. Both these stars are in the constellation of Taurus the Bull, part of a vast star-forming complex in our region of the galaxy.

The dust-system discoveries, as well as the discoveries of substellar objects around other stars, makes it even more probable that at least some other stars have planets and suggest that we are poised on the brink of discovering some of them. Every few months, it seems, the scientific journals and newspapers carry new reports that this or that group of astronomers has made the first detection of a star in the very process of forming or of a star with a possible planet or a brown dwarf. In a few cases, the astronomers or the media reporters seem to have gotten too excited. For instance, we have long known that stars such as T Tauri are young and have abundant dust, and that some even have small companions. So the actual imagery of those particular dust-disks, though interesting, is not an Earth-shaking advance. In other cases, reporters have written their stories about one star without realizing that other similar examples had already been presented (and sometimes later rejected!) in the literature of the field.

In reality, each new discovery of recent years has been more in the nature of a link added to the chain leading toward an understanding of extrasolar planetary systems rather than a climactic arrival at the end of the chain. But a growing excitement is there. The next decade will be suspenseful, as we begin to learn whether extrasolar planetary systems really exist—and in what abundance!

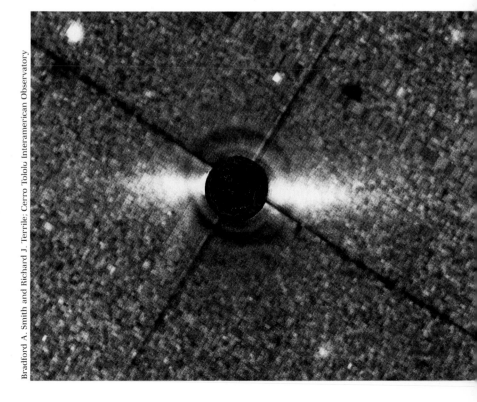

This photograph was the first direct image of a disk of gas and dust around another star. It shows an edge-on view of the disk around the star Beta Pictoris. The star itself is hidden behind a black disk at the center, inserted in the telescope to prevent the image from being "swamped" by the glare of the star. The luminous dust disk extends to either side of the star. The painting on the facing page shows what an observer near the edge of the disk might see.

Bradford A. Smith and Richard J. Terrile; Cerro Tololo Interamerican Observatory

A double sunset on an imaginary planet of a double-star system symbolizes the variety among the stars. One star is redder than the other and casts a different-color twilight glow along the horizon. The planet is located in a recently formed Pleiades-like cluster of stars, which fill the sky with light as they emerge at dusk. The brightest members of the cluster are the hottest, bluest stars, but fainter stars of other temperatures and colors can be found. Volcanic eruptions are producing this ocean-planet's only land masses, such as the dormant volcanic island on the horizon. [Preceding page]

William K. Hartmann

THE NEARBY STARS

How far away are the stars? In our model where stars were dust-motes and galaxies were continents, stars in the sun's neighborhood were a block apart. In reality, the distance to the nearest star is some 25,000,000,000,000 miles, or 25 thousand billion miles. Other stars are much farther away. This leads to such inconveniently large numbers that astronomers express the distances in special large units. A convenient one is the light-year, which is equal to the distance light travels in a year: about 5,900,000,000,000 miles, or 5.9 thousand billion miles. An astronomer, therefore, says that the nearest star is about 4.3 light-years away.

This is Alpha Centauri, the brightest star in the southern constellation of the Centaur. Not readily visible from the continental United States but well seen from Hawaii, Alpha Centauri turns out to be a triple-star system. The brighter pair, Alpha Centauri A and B, are two rather sunlike stars a little farther apart than the sun and Uranus.

The faint companion, Alpha Centauri C, is a low-mass red star too faint to be prominent in our sky and is about 400 times as far from the AB pair. It turns out to be closest of the three to us and is thus sometimes called Proxima Centauri.

Next closest after Alpha Centauri is Barnard's Star, which is 6.0 light-years away. This star is notorious for the controversy about a possible companion of less than Jupiter's mass, mentioned earlier. The most recent studies, however, suggest that the earlier "detection" was a result of random fluctuations in the observations, so that Barnard's Star may in fact have no detectable companions. Smaller, cooler and redder than the sun, it remains an interesting object for study both because of the controversy and because of its relative closeness.

The chart on the facing page lists the closest stars and their properties. Most of them are not familiar names because the most common stars scattered through space

THE 15 NEAREST STAR SYSTEMS
(OUT TO 11.4 LIGHT-YEARS)

STAR NAME	DISTANCE LIGHT-YEARS	MASS OF PRIMARY IN SOLAR MASSES	MASS OF COMPANION (IF ANY) IN JUPITER MASSES
Sun	0.0	1.0	1.0 + others
Alpha Centauri	4.3	1.1	900 and 100
Barnard's Star	6.0	0.15	None known (?)
Wolf 359	7.5	0.1?	None known
Lalande 21185 (= BD + 36°2147)	8.2	0.35	20 (?)
Luyten 726-8	8.4	0.10	100
Sirius	8.6	2.3	980 (white dwarf)
Ross 154	9.4	0.2?	None known
Ross 248	10.2	0.2?	None known
Epsilon Eridani	10.7	0.7?	? (suspected); dust detected
Luyten 789-6	10.8	0.1?	None known
Ross 128	10.8	0.2?	None known; dust detected
Epsilon Indi	11.2	0.7	None known
61 Cygni	11.4	0.6	600 and 80; dust detected
Tau Ceti	11.4	0.9	None known; dust suspected

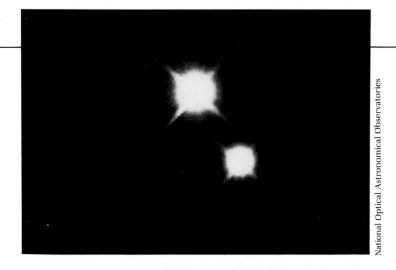

National Optical Astronomical Observatories

*P*art of the closest star system. This photograph shows the brightest two members of the famous star system Alpha Centauri. A third, much fainter member, outside this field of view, is somewhat closer to us and is called Proxima Centauri. It orbits around these two stars, called Alpha Centauri A and B, which in turn orbit around each other.

are somewhat less massive and hence less bright than the sun, and they do not look prominent in our sky even though they're close. The stars that are prominent in the sky are, in most cases, massive, superluminous stars that can be seen a long way off—"the whales among the fishes," as one early astronomer called them. But even among the nearby stars we find a few familiar names. Alpha Centauri is one, and Sirius, the star of greatest apparent brilliance as seen from the solar system, is another. It is 8.6 light-years away. Vega and Beta Pictoris, two of the stars recently found to have orbiting systems of dust particles, are twenty-six and fifty light-years away, respectively.

Since we know our sun has planets, astronomers naturally tend to favor stars similar to the sun when it comes to the search for planets. Good candidates, therefore, are hydrogen-burning stars of about one solar mass. These are not uncommon. Among the closest stars, Alpha Centauri A and B, Epsilon Eridani (a companion is suspected), 61 Cygni, Epsilon Indi and Tau Ceti are frequently mentioned in discussions of the search for extrasolar planets. These are all within a mere twelve light-years from Earth. In 1986 astronomers at Kitt Peak National Observatory and the University of Arizona concluded a search for dust among

Ron Miller

$\mathbb{S}$*irius, a relatively close star about nine light-years away, is the brightest star in our sky. It is an interesting system; the bright star, somewhat bluer, hotter and more massive than the sun, is circled by a very faint white dwarf. The above scene could result if a hypothetical ringed planet were also in the system. Here the planet is shown eclipsing Sirius, making it possible to see the faint white dwarf in the distance at the upper right. The reddened light passing through the planet's atmosphere causes red fringes at the edge of the planet's shadow, which is cast on the ring system. A satellite of the planet is also visible in crescent-phase illumination, to the right of the planet's rings.*

the nearest fifty star systems and found dust near several of them, including Epsilon Eridani, 61 Cygni and Tau Ceti. The dust is at least indicative of the raw material for planet formation; it suggests, but does not prove, the presence of larger bodies such as asteroids, comets or full-fledged planets. However, the 1987 Canadian results, mentioned in the last section, listed a small brown dwarf of two to five Jupiter masses in orbit around Epsilon Eridani.

Perhaps it is near one of these nearby stars that a full-fledged planet will first be confirmed. One of these names will then leap into prominence as the possible locale of an alien harbor for life . . .

BETWEEN THE STARS

The distance from Earth to the center of our galaxy is something like 30,000 light-years. This means, for one thing, that the light from the center of our galaxy has been traveling for 30,000 years to get here. We know only what was happening there 30,000 years ago, not today. If catastrophic explosions happen there, as many astronomers suspect, and if one happened last month, we humans would not know about it for another 30,000 years.

Our own local region, of a few thousand light-years in any direction from the sun, is a typical small neighborhood in a typical galaxy.

If we look at this neighborhood or other similar parts of the galaxy, we will find that the spaces between stars are not empty. In various places are patches of glowing fog called *nebulae,** from the Latin word for "cloud." Further study reveals that diffuse gas and scattered microscopic grains of dust pervade all of interstellar space and that the visible, glowing clouds are only local concentrations.

The appearances of the various clouds are amazingly varied. Traditional science starts by categorizing things, and astronomers of past generations developed a host of names for different kinds of clouds. *Reflection nebulae* were those whose glow was simply light reflected off the cloud from a nearby star, like a cumulus cloud at sunset on our own Earth. *Emission nebulae* were those whose glow was emitted by the cloud gas itself because its atoms had been stimulated by radiation from a nearby star, just as the atoms of gas in a neon-light tube glow as a result of electrical stimulation. Some of the old names were misleading. Certain spheroidal clouds of gas blown out by supernova explosions resembled disks of planets in early telescopes and hence came to be called *planetary nebulae,* although they have nothing to do with planets.

Today's astrophysicists are more interested in physical processes and origins of the clouds than in pigeonholing them into categories. Interstellar material has an extraordinary life cycle of its own, something of a chicken-and-egg affair, so it's hard to know where to start. Gas collapses to

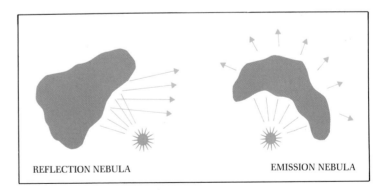

REFLECTION NEBULA   EMISSION NEBULA

A reflection nebula shines by the light it reflects from a nearby star. It has no light of its own (left). An emission nebula (right) has its gas excited by radiation from a star, which causes the gas to glow. An emission nebula is visible by its own light.

---

*We are not talking about the cocoon nebulae discussed earlier, because these are not nebulae enshrouding a newly formed star. These are independent clouds, sometimes alone, sometimes near new stars, sometimes blown out of old-age stars. Also, as noted before, they are not quite like clouds in our sky; some are more ragged, like clouds of cigarette smoke; some are more symmetric, as we will see.

form a star. If the star has low mass, it "burns" for a long time and little of the gas ever escapes back into space; but if the star is massive, it may explode as a supernova. Gas is blasted *back* into space. But the gas is altered. Some of it has been "burned" inside the star. As we saw earlier, this means that thermonuclear fusion reactions have created many atoms of heavy elements: carbon, iron and others. When the supernova star spills its insides back into space, it creates new interstellar gas that is much richer in heavy elements than the original gas that made the star. In the region of a supernova, the outrushing cloud of gas may hit surrounding diffuse gas, pushing it aside and compressing it. This may help form some new, localized, denser-than-average clouds. Some of these clouds may be so compressed that a new cycle of collapse is started. Each generation of stars may include some that blow up, add new matter to interstellar gas, trigger new collapses, start yet another generation of stars, and so on, chicken and egg.

Stars usually form not one at a time but in clusters such as the Pleiades, or "Seven Sisters," a star grouping that graces our winter skies. Because stars form in clusters, the gas-cycling processes happen not in isolation but in related sequences. A huge mass of dusty gas debris hangs around a newly formed cluster, left over from the collapse that formed the cluster itself. The first supernova blows new material into it and disturbs it. Other massive stars explode in sequence, depending on when they formed and how fast they burned their nuclear fuel. These disturbances initiate localized collapses on the *outskirts* of the cloud, perhaps a million years after the first collapse. The new collapses eventually produce some massive stars that themselves explode, disturbing gas even farther from the original cluster. Gradually, the star-forming activity may migrate from the central regions of the cluster to the surrounding volume of space. The cluster, in its orbital journey around the galaxy's center every 150 million years or so, may run into other clouds, causing compressions and triggering spreading star formation in adjacent areas. Star formation is contagious.

Some of the most interesting clouds are the densest— huge black monsters that lurk in space, almost invisible except for the fact that they block the light of stars behind them. One fairly well-defined cloud silhouetted against the Milky Way is visible from southern latitudes (Hawaii and

The well-known star Altair, about sixteen light-years away, is remarkable for its extremely fast rotation. In contrast to our sun, which rotates in twenty-five days, Altair rotates once every six and one-half hours. As a result, this bluish-white star is extremely flattened in shape by centrifugal forces. Here it is seen from the polar mountains of a hypothetical nearby planet. A large moon of the planet hangs over the mountain range.

Ron Miller

southward) and is called the Coal Sack. In the northern hemisphere, silhouetted against the Milky Way in the constellation Cygnus, we can see more diffuse dark clouds that are sometimes called the Great Rift because they split the Milky Way into two branches.

THE CHANGING COMPOSITION OF INTERSTELLAR GAS

As each supernova explodes, it blows out gas rich in the heavy elements formed inside the star. These elements, of course, are immediately diluted by the hydrogen and helium that make up 97 percent of the surrounding interstellar gas. Nonetheless, as time goes on, this process means that interstellar gas gets richer and richer in elements such as carbon, silicon and iron. The upshot is that the interstellar gas of our whole galaxy is accumulating elements that make dust grains and planets; our galaxy's planet-spawning ability is increasing!

Just as interesting to astronomers is the converse implication: the early gas of our galaxy must have been much poorer in heavy elements. It must have been nearly pure hydrogen and helium. As we will see, astronomers have confirmed this by direct observation. The oldest stars are indeed nearly pure hydrogen and helium.

Supernovae produce expanding bubbles of hot gas blown off their outer layers. The atoms and electrons of this gas are disturbed and glow with various colors. From a distance, most of these bubbles look like diffuse rings or doughnuts, since our line of sight passes through more gas along the edge of the bubble than through its center. Such nebulae were long ago named "planetary nebulae" because they looked like faint planetary disks through early telescopes. In reality, they have nothing to do with planets. The supernova remnant star resides in the disk's center. The nebula is viewed here from the surface of a planet whose two moons are also visible. The planet's sun is offstage, having just set below the horizon to our left.

MOLECULAR CLOUDS: POTENTIAL FOR LIFE

The densest clouds are called *molecular clouds* because they contain concentrations not only of single atoms but also of atoms joined together into molecules. Walt Whitman once said that every cubic inch of space is a miracle. A cubic inch of a molecular cloud is a miracle because molecules are miraculous in space. Decades ago, astronomers calculated that atoms in the *average* interstellar region were so far apart that collisions—needed to build up molecules—would be very rare. And the gas inside stars is so hot that the collisions are too violent to build up molecules; on the contrary, they break molecules (see Chart on page 25). But inside the cooled, collapsing, dense

clouds, atoms collide and combine into molecules. Astronomers were thus not too surprised to discover a few simple molecules in space, such as oxygen and hydrogen joined to make OH, the so-called hydroxyl molecule. They were amazed, however, as results began to pour in from infrared telescopes in the 1970s: there were not only simple molecules, but also incredibly complex molecules in dense clouds. Not only 3-atom molecules like water (H_2O) and hydrogen sulfide (the rotten-egg-smelling gas, H_2S), or 4-atom molecules like ammonia (NH_3), but also 11-atom whoppers like cyano-octa-tetrayne (HC_9N)!

We are talking about big organic molecules, based on carbon atoms and having the ability to link into long chains. Amazingly, astronomers found that molecular clouds contain many of the big, complex molecules that are the building blocks of life. These include amino acid molecules, formaldehyde (H_2CO, the liquid used to preserve biological specimens), formic acid (HCOOH), methanol (CH_3OH), vinyl cyanide (H_2CCHCN), and others.

These molecules are not the same as living material, of course, and they do not prove that life has existed elsewhere. They were formed by the simple, nonbiological process of atoms hitting each other and sticking together. But they show that the universe is pregnant with possibilities for life.

Some scientists have speculated from these findings that interstellar organic molecules, created in vast molecular clouds before the formation of our solar system, impregnated the icy comets and dust of our solar system and became the ultimate sources of Earth's organic material.

Could it have been interstellar molecules from which life grew on Earth after comets and meteorites crashed into the planet and delivered their fertilizing cargos? Joni Mitchell may have been right in her song "Woodstock": perhaps "We *are* stardust," not only in the sense that our carbon and oxygen atoms were synthesized by nuclear reactions inside stars but also in the sense that our ancestral organic molecules came from starry molecular clouds. This last idea, however, contrasts with the more widely accepted theory that organic materials synthesized so fast by natural processes in the stormy oceans of Earth that any input from the sky was negligible.

The famous English astrotheorist Fred Hoyle and his colleague Chandra Wickramasinghe have gone far beyond these ideas. They speculate that virus molecules could grow in molecular clouds and be trapped in comets and that these fall on the Earth today, initiating epidemics and explaining why certain viral diseases seem to appear suddenly, rampage through the population and then die out. These ideas were developed in their book *Diseases from Space* (1979). Moreover, they speculate that full-fledged bacteria might have grown in space and compare the spectra of interstellar clouds with the spectra of light passed through bacteria, cellulose and other such substances. Most astronomers and biologists, however, remain unconvinced that evolution has gone so far in interstellar space. They argue that the spectra reveal only the richness of nonbiological, organic materials in interstellar clouds and that disease-causing viruses, as well as terrestrial life itself, are a product of Earth.

M*any stars, and hence possibly many planetary systems, are closer to vast nebulae than we are. Here we see the view from a crater rim on a moonlike world near a huge nebula and molecular cloud complex, resembling the Orion nebula. The planet circles an offstage massive, young, violet star. Perhaps this star itself formed recently in the distant cluster and is now escaping from it, moving with its planet toward deep interstellar space.*

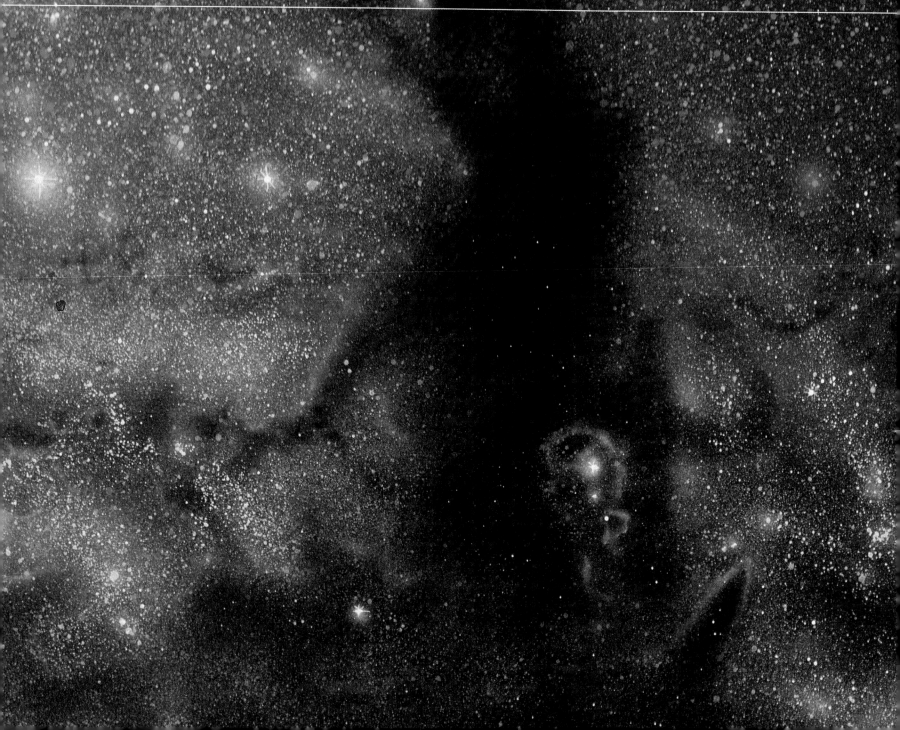

Detailed studies of photographs of various spiral galaxies reveal ragged and chaotic distributions of dust and stars along their central planes. While most of the dust lies along the galactic plane, shown here running left to right, occasional columns, such as the bright column to the right, seem to extend "upward" or "downward." These may be initiated by violent events such as supernova explosions or gravitational disturbances of the dust by passing star clusters. In the distance (at left) are the reddish stars of the galactic hub, partly hidden behind dust clouds. [Preceding page]

Ron Miller

THE NATURE OF THE MILKY WAY

The galaxy in which we live is shaped like a phonograph record with a central bulge, or perhaps more like two dinner plates face to face, also with a central bulge. We live partway out in one of a set of ragged, spiral arms, winding out from the bulge. Interstellar clouds and star clusters are the main features of the spiral arms, strewn along them like beads on a string.

According to recent theories, the arms are somewhat related to the spiral-galaxy-shaped pattern you can generate in your coffee cup by stirring the coffee and dropping cream into it. As in the coffee cup, the central regions of the galaxy are orbiting around its center faster than the outlying regions—a consequence of the laws of orbital motion. This effect alone partly explains the spiral arms: if you pour a teaspoon of cream into the coffee all at once, the rotational shear creates spiral windings. But this is not the whole story. The galaxy is old enough for the inner part to have made about fifty rotations. If you stirred your

coffee fifty times, the cream windings would be so tight that you'd have trouble detecting any spiral pattern. Similarly, if you want to believe that galaxies' spiral arms come merely from rotational shear, you have to admit that fifty rotations of our galaxy would wind the arms too tightly for us to detect them. Instead, the arms of our galaxy and most others are loose, often showing less than one complete turn.

The explanation is that the spiral arms are defined by clusters and nebulae that have a much shorter lifetime than the age of the galaxy. As explained earlier, clusters form, initiate bursts of star formation in the neighborhood and eventually disperse as the star-forming material dies out and the stars slowly move away from each other. So each of the brightest groups of massive, newly formed stars, which catch our attention and help us define a pattern when we look at a galaxy, is prominent for perhaps only 100 million years. Since the material takes about 230 million years to move around our galaxy, at Earth's distance from the center, star-forming regions in our neighborhood are sheared out into a spiral arm, making perhaps half a turn around the center. A better coffee cup analogy would be to stir the coffee, then dribble only a drop at a time into the coffee at different distances from the center (and use an eyedropper if you want to be really scientific). Each drop represents the flare-up of a new burst of star formation. Due to the faster rotation toward the center, each splotch of cream (a group of new stars) shears into a spiral arm.

The features of the galaxy in which we live are not just theoretical patterns drawn on some theoretician's computer screen. The geometry of our galaxy directly affects the appearance of our night sky. From inside the galaxy, we survey the night sky and see a hazy band of light encircling it: the Milky Way. It is our galaxy seen edge-on—just as an ant trapped inside two facing dinner plates could look around and see the surrounding rims.

As Galileo discovered when he used the first astronomical telescope in the 1600s, the Milky Way is really a mass of thousands of stars, star upon distant star, unresolved individually by the naked eye. Don't expect to see it from your city backyard or apartment balcony. You must go far into the country, away from city light pollution, to see a truly dark sky where the galaxy stands out. Wait five minutes after stepping outside, so that your eyes become adapted to the dark. On a clear summer or winter evening, you will discern this hazy, glowing swath across the sky. In the summer, if you look into the southern sky near the constellations Scorpio and Sagittarius, you're looking toward the center of the galaxy. The Milky Way is somewhat brighter there, but the massive dust clouds between us and the center obscure a direct view of the central bulge. These clouds make dark splotches and rifts in the softly glowing milky band.

On a winter night, if you look toward Orion, you're looking down one length of our spiral arm. Again, dust clouds obscure a very distant view, but if you scan back and forth along the Milky Way in this part of the sky, you'll notice that the Orion region marks a concentration of bright stars. This is because in this direction you're looking down the length of our spiral arm, and star formation has been happening there, especially in a region about 1,300

William K. Hartmann.

Cerro Tololo Interamerican Observatory

A comparison between our galaxy and another, distant galaxy. When we look at the Milky Way (top), we are seeing an edge-on view through our own galaxy. In telescopic views of certain distant edge-on galaxies (such as one cataloged by astronomers as NGC 4565, seen at bottom), we see similar features. These features include a bulging bright hub forming the central region (lower left in each image) and dark clouds of dust blocking the light of some of the stars in the galactic disk. The top photo is made with an ordinary 35mm camera with a fish-eye lens, covering about 120 degrees across the sky; it is a 30-minute exposure at f2.8 on 2475 Recording Film.

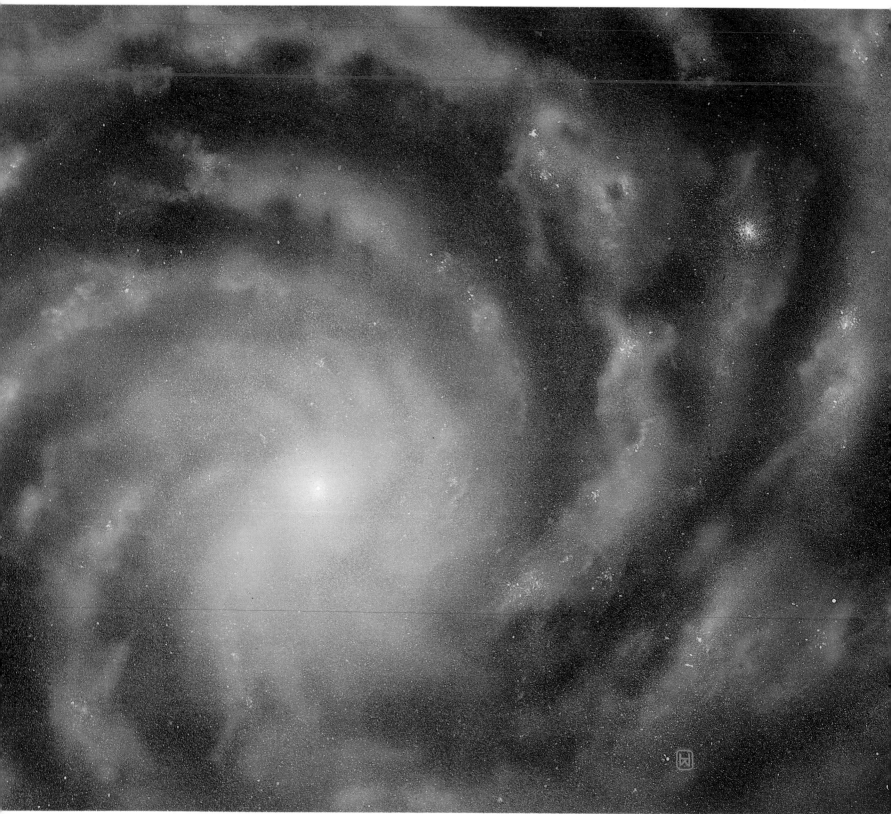

William K. Hartmann

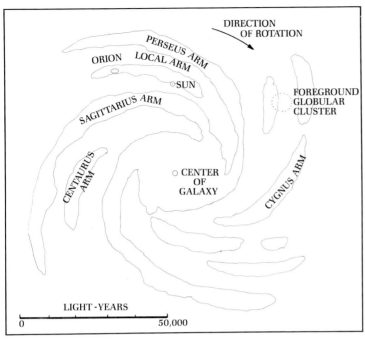

A view of the Milky Way (far left) from "above" the galactic plane. The position of the spiral arms and central region is shown as mapped by radio astronomers, who can measure distances to clouds of hydrogen gas by their radio emissions, and by optical astronomers, who can measure distances to various star clusters. The central regions are more ruddy in color due to their preponderance of "red giant" stars. Our sun would be one tiny dot in one of the upper, central arms. The key at the near left shows some of the known features; local spiral arms are named for prominent constellations of stars that lie in their direction.

light-years away in the heart of the constellation Orion. The central "star" in Orion's belt, if examined with binoculars or a small telescope, turns out to be a large, glowing nebula. Studies of this region have revealed a hotbed of star-forming activity; bright, young stars form much of Orion and dot the surrounding sky.

Many people are surprised that these awesome features of our galactic environment can be seen with the naked eye and are not just something for astronomers with million-dollar observatories. One need only understand *what* one is seeing, and then the story of the sky begins to unfold. The cosmic show surrounds us—if we know what to look for. As Henry Thoreau said, "There is just as much beauty visible to us in the landscape as we are prepared to appreciate, not a grain more." In this case, the "landscape" is cosmic.

GLOBULAR CLUSTERS

Astronomers have discovered clusters of stars that lie outside the Milky Way in our night sky and hence outside the spiral arms. Triangulation in three dimensions shows that these clusters are scattered through a spheroidal volume of space surrounding our disk-shaped galaxy, "above" and "below" its plane. They are like fuzzballs of cotton placed above and below our dinner plates. They look much different from the star-forming clusters in our spiral arms, which are loose accumulations of a few hundred stars each. *Globular clusters*, as they are called, are tightly packed spheroidal groupings of as many as a million or more stars.

Globular clusters orbit around our galaxy and occasionally, therefore, pass through the plane of the galaxy and out the other side. They are too far away from the solar system to be prominent to the naked eye, although an amateur's telescope reveals them impressively. But one

Ron Miller

Some large spiral galaxies, including our own, have small satellite galaxies that contain the stars and dust necessary to spawn planets. One such imaginary planet here offers an extraordinary nighttime sky, nearly filled with the face-on view of our Milky Way's galactic spiral and bright central nucleus. Imagine the mythology and astronomy that might develop among sentient creatures living on a planet with a glowing, celestial spiral illuminating its nights.

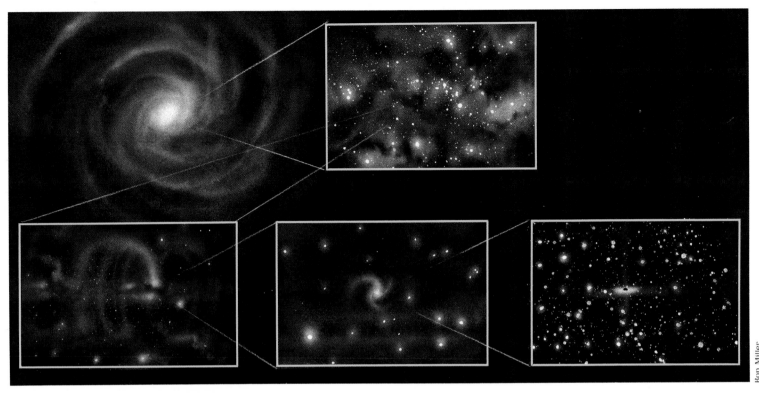

Ron Miller

Zooming into the center of the Milky Way. This series of drawings gives an idea of a journey into the heart of our galaxy. Step 1 (upper left) shows the whole Milky Way structure, with its spiral arms. In step 2 (upper right) we zoom from our spiral arm toward the center, seeing a field of dust clouds inside one of the inner spiral arms. In step 3 (bottom left) we approach the mysterious central region, packed with red giant stars and energetic gas wisps. Step 4 (bottom center) shows a spiral structure detected by radio astronomers; it may be gas spiraling into the nucleus. Step 5 (bottom right) shows the central accretion disk that may surround a giant black hole at the heart of the nucleus.

that happened to pass near us—say, at the distance of the Orion region—would light up our night sky with a fantastic clumping of thousands of stars comparable to the handful of brightest stars in our present sky.

By measuring the evolutionary time required to produce the large numbers of red giants observed in globular clusters, it has been possible to calculate their ages. Most of them formed an estimated 12 or 13 billion years ago, which has been interpreted to be the *age of the Milky Way galaxy itself*. According to our current understanding, the Milky Way formed when a vast cloud of primordial gas began to collapse. As it shrank, subclouds collapsed within it; these in turn subfragmented into stars, forming the globular cluster. Since the globulars formed so early, they were left "stranded" in a spheroidal swarm surrounding the still collapsing Milky Way. Because of its random rotational motions, the pre-Milky Way cloud had angular momentum and flattened into a spinning disk, a gigantic version of the flattened cocoon nebulae surrounding the primordial sun and other young stars. Further subfragmentation within this disk led to the clusters and spiral arms that characterize our galaxy today.

Ron Miller

Although planets are unlikely to form inside globular clusters because these clusters lack heavy elements, close-up views of globular clusters from planets may occur. Globulars, in their orbital trips around the centers of galaxies, pass through galactic disks, where the heavy-element-rich stars and dust (and hence, we assume, planets) reside. For about 100,000 years, this globular cluster will dominate the cloudy sky of this gas giant planet as the cluster passes through the galactic plane near the planet.

"POPULATIONS" OF STARS

As astronomers mapped stars and their characteristics in different parts of our galaxy, an extraordinary fact emerged. Stars in the central bulge differ in composition from stars in our neighborhood. This is revealed by observations with the spectrometer, which shows that within the star light of certain colors is absorbed by different elements. One set of absorptions, or missing colors, reveals hydrogen; another iron; and so on. As remarked earlier, stars near the sun consist of about 97 percent hydrogen and helium gases, with only 3 percent being the heavier elements such as carbon, oxygen, silicon, iron, etc. But the stars in the hub of the galaxy are composed almost entirely of hydrogen and helium. All the heavier elements comprise perhaps only a tenth of one percent of the material!

The globular clusters, too, lack heavy elements and seem to be the same types of stars as in the central bulge. In fact, the central bulge of a galaxy like ours bears some resemblance to a giant globular cluster.

Astronomers use the term "population" to describe the different compositional types. There is a heavy-element-poor population near the center and in the globular clusters, but a heavy-element-rich population out in the spiral arms. What can this mean? We saw that heavy elements must be accumulating in star material because they are being formed continually inside massive stars and blown into space. So the discovery that the population of stars in the hub has almost no heavy elements means that those stars were formed long ago and that the primordial gas of the galaxy must have been nearly pure hydrogen and helium.

There are several ramifications to this discovery. First,

it strongly supports the whole theory of star evolution and the chicken/egg processing of interstellar material in and out of stars.

Second, it predicts that there should be very little dust among the heavy-element-poor population, because the dust is made from heavy elements such as carbon, silicon and oxygen. This is confirmed by observation: the hub region and the globular clusters are generally poor in dust, although clouds of dust can be found near the galactic center, where giant stars may have exploded. The lack of interstellar dust clouds, in turn, inhibits new star formation, so that the whole star-forming and star-cycling process seems to be retarded in the heavy-element-poor populations. Only a single generation of stars, more or less, formed at the beginning in these regions, while in the spiral arms new generations of stars are forming continually. Throughout much of the swollen central hub of our galaxy, star formation ended long ago.

Third, it means that there may be few or no planetary systems in these regions. Our planet is made of heavy elements, especially oxygen, silicon, magnesium, aluminum, and so on. Even Jupiter, which has a core of ices and rock, must have needed icy planetesimals containing oxygen in order to start forming. In much of the galactic hub and in the globular star clusters, the lack of ice- and rock-forming elements may have prevented the formation of planets as we understand them.

One can always hope that on a trip through the galactic disk a globular cluster might have captured a planetary system or some individual planet. Imagine the view from a planet inside a globular cluster: a landscape with a million stars as bright as Sirius scattered across the sky!

Ron Miller

Two views, eons apart, from a planet on the edge of the hub of a spiral galaxy. In the first view (top), shortly after the galaxy's formation, the central hub stars are newly formed. No heavy elements or dust have yet been ejected to cloud interstellar space. The brightest stars in the hub are young, hot, blue stars. In the second view (bottom), the galaxy's stars have evolved and supernova outbursts have ejected gas rich in heavy elements. This material clouds interstellar space with "smog" clouds of dust and large molecules. The brightest stars in the hub have evolved into red giants. Meanwhile, the imaginary foreground planet has evolved as well; the planet has a very thin atmosphere, and meteorite bombardment and other processes have weathered the distant peaks. The moon pictured in the first view has spiraled inward toward the planet and broken up due to tidal forces, creating a ring system of debris resembling an immense, colorless rainbow.

THE GALACTIC NUCLEUS: THE MONSTER IN THE MIDDLE

Some 30,000 light-years in the direction of Sagittarius is . . . well, *something*. As early as 1918, astronomers realized that the center of the galaxy's disk lay in the direction of the constellation Sagittarius. But because of the dust clouds, no one could see what was there. The advance that allowed us to begin to probe the nature of the galactic center was the invention of radio and infrared astronomy, which allows us to "see" through the dust clouds and map the distribution of material at great distances. To put it more scientifically, radio waves and infrared waves can pass through large amounts of material (as you know since your radio can pick up a radio station from inside your house); therefore, we can pick up the strong radio and infrared emissions from the galactic center, beyond the interstellar dust clouds.

Even the very earliest radio astronomy revealed that something remarkable lay beyond the dust clouds in the direction of the center. One of the first radio telescopes—a large antenna intended to pick up radio waves from the sky instead of from radio stations—was built in the 1940s by amateur astronomer and radio buff Grote Reber in his backyard. Reber began mapping the sources of celestial radio noise and showed that the strongest radio emission comes from the direction of the galactic center. In the 1950s, '60s and '70s, radio astronomy and infrared astronomy were used to map the structure of the spiral arms and to detect an outward-moving spiral arm of hydrogen gas some 10,000 light-years from the center, or about a third the distance of our own arm. Eventually, astronomers de-

tected the small, intense source of infrared and radio emission that marked the true nucleus of the galaxy. Its visible radiation is hidden by intervening dust clouds.

At the same time, photographs of many other galaxies revealed that they have very small but very bright central objects. These nuclei are usually directly visible to us because we look "down" at an angle into their centers and our view is not blocked by intervening dust. The brightness of the nuclei varies from galaxy to galaxy. Usually, they are the brightest object in the galaxy, but some are far brighter than others, giving off extremely strong radio signals as well as other forms of electromagnetic radiation.

Many astronomers try to take longer and longer exposures of galaxies, and invent sensitive devices to get images of the faintest, outlying gas around the edges of the galaxies in order to understand their shape and dynamics. But if shorter and shorter exposures are taken, the faint outer regions disappear, leaving a single, starlike image. It is a mysterious pinpoint of light, the bright *something* in the center—the enigmatic nucleus! This is a dramatic proof of how important the nucleus is, compared to the much fainter outer regions.

In some galaxies, splattered streamers of hydrogen gas extend out from the nucleus, apparently ejected explosively. This means that some energetic process is disturbing the gas in these regions, causing explosive outbursts of energy. A number of galaxies have vast clouds of hydrogen placed almost symmetrically above and below the galactic plane, apparently ejected in each direction from the nucleus. In some of these, we can see bizarre, narrow jets shooting directly out of the nucleus in the "upward" and "downward" directions perpendicular to the plane. This phenomenon is called *bipolar outflow*. Some process in the nucleus is focusing gas into two beams and squirting it out from the galaxy, like a garden hose with two nozzles aimed in opposite directions.

The galactic nucleus closest to us, which ought to be the one we could study most carefully, is the one in our own galaxy. It is less directly visible to us because of the cosmic cloudbank of the Milky Way's inner spiral arms. Two approaches are available to study it, and both are limited in their rate of progress more by the amount of research funding available around the world than by scientific factors. The first is theoretical research into the processes that might cause high-energy events in the center of our galaxy or of others. What might lead to the expulsion of hydrogen or the flare-up of the central region, making it outshine all the other stars of the galaxy put together? The second is the observational pursuit of ever more detailed images, either at radio wavelengths or at infrared wavelengths that would pierce the intervening dust.

Theoretical work has promoted the idea that, in the crowded central regions of our own and other galaxies, collisions of stars or inward spiraling of gas onto a giant central star may have led to formation of one or more huge objects, perhaps hundreds or thousands of solar masses. These would have burned their nuclear fuels explosively, collapsing rapidly to form a black hole. Subsequently, some of the gas ejected from nearby red giants and supernovae would be attracted toward the black hole by its enormous gravity. Because of the initial orbital motions of the gas in the rotating galaxy, modified by collisions with other gas clouds in the neighborhood of the black hole, the gas approaching a black hole is expected to accumulate in a disk-shaped system around the hole and slowly spiral inward. Such a disk of gas and dust is called an *accretion disk*, because it is composed of material accumulating, or accreting, onto the central object. Losing orbital momentum due to the friction of collisions, more and more material could continually spiral onto the inner part of the disk and eventually onto the black hole.

Because new, incoming gas from nearby supernovae would fall onto the accretion disk with immense speed, the accretion disk would be extremely hot. According to theoretical models, it could radiate not only bluish visible radiation but also ultraviolet, X-ray and gamma-ray radiation. Although radiation cannot escape from inside the so-called event horizon of the black hole itself, the disk would be outside the event horizon and this unusual radiation could escape. At greater distances from the disk, gas clouds might be near, or somewhat warmer than, normal interstellar temperatures and hence would emit the usual infrared and radio radiation associated with such clouds. These ideas explain some of the kinds of radiation observed to come from galaxies' central cores.

Some of the gas spiraling through the accretion disk would be accelerated by intense, twisted magnetic fields around the black hole. Escape of the gas in the equatorial plane would be blocked by the surrounding accretion disk, but as gas pressure increased, some gas might be squirted out along the "north" and "south" polar axes of the system. The theory is woefully inadequate at this point, but further study along these lines might explain the remarkable bipolar jets of hot gas shooting out of some galaxies.

Meanwhile, observers are using new infrared equipment and new arrays of radio telescopes that can resolve extremely fine angular details in celestial radio sources. Recently, they have mapped new details of the galactic center. The outward-rushing spiral arm near the center turns out to contain about 10 million solar masses of solar hydrogen and, judging from its rate of motion, was pushed outward perhaps 10 or 100 million years ago. In cosmic terms, this is a surprisingly short time—only a fraction of one percent of the age of the galaxy! So some "recent" energetic events have expelled material from the nucleus' vicinity, supporting the idea of explosive events in galactic nuclei.

Mapping areas closer to the center, observers found that the region about 1,000 light-years wide around the nucleus contains huge numbers of red giant stars as well as clouds of dust and gas. Radiation from the gas seems to be the type that comes from interactions of electrons with strong magnetic fields. This ties in with the idea that the nucleus is surrounded by the strong magnetic fields that would be expected from a collapsed object.

Within a few hundred light-years of the center are 60 million stars and clouds of ionized hydrogen, along with

Ron Miller

Once upon an imaginary time, a globular cluster's orbit made it collide with a star and its planetary system. As the star passed through the tightly packed cluster, repeated gravitational interactions with the cluster's stars disrupted the planetary system. One planet was ejected from an orbit around the star and captured in an orbit bound to the cluster. As the cluster continued on its 200-million-year orbit into deep space beyond the parent galaxy's disk, the planet was carried along, wandering on the cluster's outskirts. Here we see the sky of this lonely planet, a sky with a million-star constellation: the thief-globular cluster. Robbed of a nearby sun, the atmosphere of this once vibrant planet has frozen into oceans of ageless ice.

A t the heart of our Milky Way. In the distance at right is the bright "monster in the middle," the actual nucleus of our galaxy. Observations in the mid-1980s revealed adjacent light-year-thick streamers of gas in hooplike filaments, probably oriented by the strong magnetic fields of the nucleus. The nucleus itself is hidden behind a disk of gas and dust, here seen edge-on. It is probably an enormous black hole, perhaps larger than the sun but smaller in dimension than our solar system. In the nearby regions surrounding the nucleus and its filaments are dust clouds and a vast number of stars whose brightest examples are red giants. The dust in this region is an extension of the dust lying throughout the plane of our Milky Way. But the rest of the central hub of our galaxy, above and below the plane, is relatively dust-free.

Ron Miller

A jetting galaxy seen in the sky of a moonlit, icy world located in a nearby satellite galaxy. The sky of this imaginary world is dominated by a large, barred spiral galaxy. The nucleus of the barred galaxy is active and possibly contains an accretion disk around a black hole. By processes still uncertain to astronomers, the nuclei of galaxies sometimes produce high-speed jets of gas nearly perpendicular to the galactic system.

dust. Some gas seems to be organized in filaments a few light-years wide, stretching over arcs perpendicular to the galactic plane and about 100 light-years long. These look like a pattern organized by magnetic fields, again supporting the presence of magnetic effects.

Within ten light-years radio images show a crude spiral with three main arms, possibly material spiraling in toward the central nucleus, which is still much smaller. It's amazing to think that features of galactic scale—50,000 light-years in dimension, such as the bipolar jets of some galaxies—could be spawned by something less than a few light-years across! With this in mind, we keep probing toward smaller and smaller scales.

Within a few light-years of the center is a mind-boggling swarm of a few million stars, along with dust and gas. Here is an environment worthy of our imagination! By comparison there are 10,000 stars in a comparable volume in the center of a typical globular cluster, and only one in a comparable volume near the sun. The separation distance between stars would be only 2 percent of a light-year, or perhaps thirty times the distance to Pluto. Many of the stars would be red giants, and the sky here would be packed with hundreds of stars as bright as the full moon, averaging only about five degrees apart.

Gamma ray emission from this region is consistent with theoretical models of gamma ray emission from colliding subatomic particles boiling off the hot accretion disk. Detail at finer scale is difficult to obtain. High-resolution radio astronomy studies, combining data from several telescopes, suggest that there may be an object at the center of the complex about the size of our solar system, perhaps no more than the diameter of Jupiter's orbit. It may be the accretion disk itself, surrounding the black hole.

What is the role of this central object, be it a black hole or a hitherto unknown type of superstar? Is it constantly growing by swallowing gas and debris blown out of stars in the center? Do stars occasionally collide with it and produce outbursts? How often does it display explosive activity? Will it produce bipolar jets like those seen in some other galaxies? Research on the nucleus of our galaxy and those of other galaxies may answer some of these questions. In any event, this research will remain an exciting frontier of astronomy over the next decade or more.

A UNIVERSE OF GALAX

A grand view of the universe from the outskirts of the Milky Way. In the left foreground is a planet, formed from the dust of a newborn star. New stars are forming in the red-glowing nebula at bottom right, and a nearby hot, young star radiates bluish light in the bottom center foreground. In the middle distance are the dusty arms of the Milky Way—the spiral galaxy in which we are located. The dominant color is bluish, from the light of the brightest, hottest, newly formed stars. In the distance is the orangish glow of the galactic hub where star formation has ended and the brightest stars are old red giants. Scattered above the plane of the Milky Way are globular clusters, orbiting around the galaxy. In the far background, here and there, are distant galaxies, which contain their own stars—and perhaps alien planetary systems. [Preceding page]

Pamela Lee

DISTANT GALAXIES

Beyond the edges of our galaxy, strewn as far as the telescope-aided eye can see, other galaxies are scattered through space. They are surprisingly close together, in terms of the scale of their own diameters. Unlike planets, which may be several thousand of their own diameters apart, or stars, which are several million of their diameters apart, galaxies are often only ten diameters apart from each other. This is why we can easily see the three nearest galaxies with our naked eye.

The closest two, unfortunately for North American observers, are far down in the southern hemisphere and visible only in equatorial regions and farther south. They are the so-called *Magellanic Clouds*, known to early Arab astronomers and to Amerigo Vespucci, and later well reported in European literature by Magellan and his crew when they sailed through southern seas on the first voyage around the globe. They are much smaller than our galaxy and look like small, hazy clouds. They may be satellites of ours, orbiting around the Milky Way. The third is farther away but is the closest large galaxy, comparable in size and appearance to our own: the famous *Andromeda galaxy*. It can be seen by the naked eye as a small, glowing, fuzzy patch in a dark (non-urban, non-moonlit) sky, if you know where to look. Its faintness to our naked eye, similar to the faint, hazy light-level of our own Milky Way, reminds us that galaxies are never dazzlingly bright structures if seen from nearby. Only the nuclei or central regions can be extremely brilliant; the rest glows with the soft light of luminous paint on a watch dial in a cave at night.

Perhaps a word of clarification is needed. You may think that the hazy glow of the Andromeda galaxy looks faint only because it is far away. There are two visual properties we need to distinguish: *total apparent brightness* and *surface brightness*. The first is the total amount of light reaching you from the object. For example, suppose the

Our own local cluster of galaxies. The Milky Way is seen in the foreground (left) with its satellites, the irregular-shaped Magellanic Clouds. In the distance is the Andromeda spiral galaxy (top left) and a still more distant neighboring spiral (at right).

Ron Miller

luminous object being considered (e.g., some galaxy) is the only light source in the night sky. Suppose you want to try to read a newspaper by this light. You face away from the object and hold up the paper, hoping to catch enough light to read by. The amount of illumination on the paper is a measure of the total apparent brightness. The total apparent brightness of a luminous object increases as you approach it, so the newspaper becomes easier to read.

The surface brightness, on the other hand, is the amount of light per unit of angular area—say, in one square degree. If you faced the object and looked at it through a small hole cut in a piece of black velvet (a fixed-size hole, cut to equal one square degree when held at arm's length, for example), surface brightness corresponds to the light you perceived through the hole. The surface brightness of a representative part of the galaxy—a portion of the spiral arms, for example—will not increase as you approach. To put it another way, the exposure you would need to photograph a particular region would not change as you approached. In the same way, the exposure you would need to photograph a portion of the moon through the window of your spaceship, or a portion of the ground, or a drive-in movie screen at night, does not

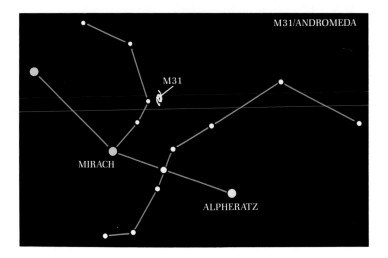

The nearby galaxy in the constellation of Andromeda can be located with the naked eye on a dark night. It appears as a diffuse, elongated disk somewhat longer than the width of the full moon.

THE FAMILY OF GALAXIES

ELLIPTICAL GALAXIES

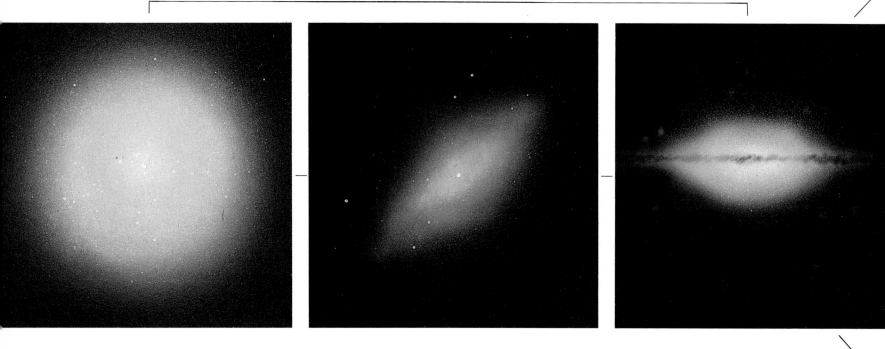

Astronomers have classified galaxies according to a "tuning fork" diagram of galaxy-types. The "handle" consists of elliptical galaxies with different degrees of flattening. It branches into two types of spirals, normal and barred. The formation of a "bar" of stars and dust in spiral shape may be a result of certain rotational properties in the protogalaxy system. Irregular galaxies have no well-defined geometric shape, but their star types seem to relate them more closely to spirals than to ellipticals. The exact explanation of the variety of types of galaxies still eludes astronomers, although initial conditions of mass, rotation rate, and age are probably involved.

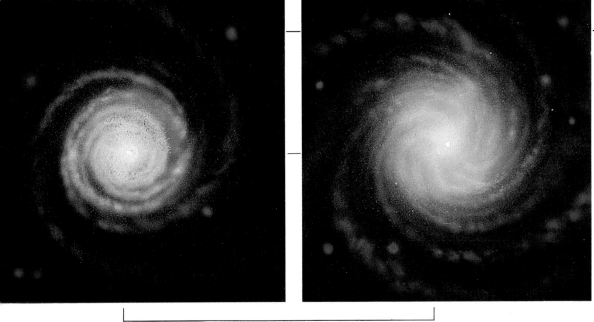

NORMAL SPIRALS

IRREGULAR GALAXIES

BARRED SPIRALS

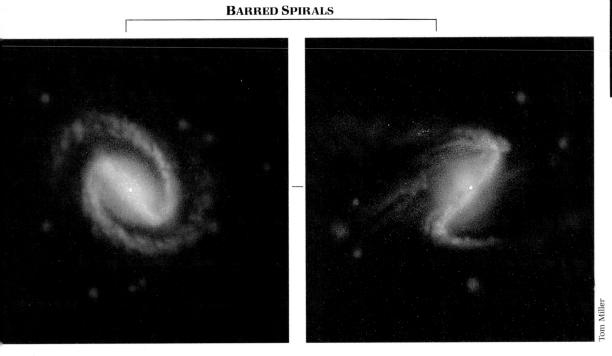

change as you approach the target. That is why the Milky Way and the Andromeda galaxy, similar spiral galaxies at vastly different distances, have roughly the same surface brightness to our eyes—a soft glow.

TYPES AND MOTIONS OF GALAXIES

There are many types of galaxies. *Spiral galaxies*, with delicate spiral arms winding out from their centers, are the type best known from popular images. There are two types: pinwheel-like spirals, like ours, and barred spirals, in which the spiral arms emanate from the two ends of a central bar-shaped swarm of stars. Initial conditions of angular momentum and gas distribution may cause some spiral galaxies to evolve toward producing "bars," but astronomers are not sure. In any case, the spiral form, characteristic of our own Milky Way and the Andromeda galaxy, is in the minority. Only about a sixth of all galaxies are spirals.

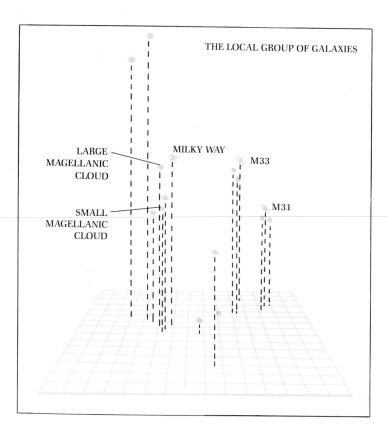

THE LOCAL GROUP OF GALAXIES

LARGE MAGELLANIC CLOUD

MILKY WAY

M33

SMALL MAGELLANIC CLOUD

M31

The surface brightness of spiral arms and other outlying regions of galaxies is quite faint, though visible to the naked eye. On a planet with an atmosphere, the faint light of evening dusk almost obliterates the features of a typical nearby spiral galaxy except for the brilliant central nucleus, which would appear like a bright, diffuse evening star. It would be intriguing and moving to watch the galaxy emerge in its full glory as the "sun" sets and the sky darkens. [Right]

About two-thirds of all galaxies are smaller assemblages of stars that look like overgrown globular clusters. They are called *dwarf elliptical galaxies*, since they are spheroidal or elliptical clumps of stars, not flat disks with spiral arms. They have no arm structure at all. They usually consist of the heavy-element-poor population of stars and have little dust. Their overall color is reddish, because their brightest stars are relatively old stars that have evolved to their red giant stage. Dwarf elliptical galaxies come in various sizes, grading up to fairly large objects that look like central bulges of spiral galaxies but without the surrounding disk of spiral arms.

Another 5 percent of galaxies are called *giant ellipticals*: they have the same shapes and properties as the dwarf ellipticals but are much larger. They have diameters as large as, or even exceeding, the diameters of the spiral disks. They are among the most massive galaxies. Another sixth of galaxies are classified as *irregular galaxies*. Their wispy shapes befit their name.

As part of the Shapley and Hubble revolutions around 1920, which proved that we live in a galaxy separate from

By measuring directions and distances of the nearby galaxies, astronomers can construct three-dimensional maps of the galactic space surrounding the Milky Way. There are two large spirals nearby (the great Andromeda galaxy, cataloged as M31, and another spiral cataloged as M33), together with a number of smaller elliptical and irregular galaxies. [Left]

Ron Miller

Ron Miller

The most common of galaxies: a dwarf elliptical galaxy (center). These are large accumulations of stars, almost like glorified globular clusters. But they are larger and often flatter than most globular clusters, though smaller than most spirals and giant ellipticals. The somewhat redder color of the globular cluster just below and to the right of the galaxy suggests that it has more red giants than the elliptical and hence may be somewhat older.

Mt. Wilson and Palomar Observatories

other galaxies, Edwin Hubble made an extraordinary discovery about galaxies. He began using the Doppler effect to measure the motions of galaxies. As you may recall from our discussion of the search for extrasolar planetary systems, the Doppler effect allows us to measure the rate of motion toward or away from us. To the complete astonishment of astronomers, Hubble found that all except the very closest galaxies are rushing away from us! And the farther away the galaxy, the faster it is rushing away!

Could this be interpreted to mean that, even after the Copernican revolution, the Darwinian revolution and the Shapley revolution, we live in the center of the universe after all? Is the Milky Way somehow "fixed" in the center, while everything else moves radially away?

No. Everything else *is* moving radially away, but this does not prove we are in a privileged position. An observer in any other galaxy would see the same thing: galaxies moving away from him or her (or it, if the observer happens to be an alien blob). To understand this, imagine that all the galaxies are corks floating in a vast ocean. Each "galaxy" has an ant living on it. The corks have been released in the ocean and are gradually dispersing from each other. Each ant looks from his cork-galaxy and observes all the other cork-galaxies and ants moving away from him. Is any one ant in some magically fixed position, as measured on some absolute scale? He would really have to measure such an "absolute" fixedness relative to a latitude and longitude position in the ocean. But how could he do this? The ocean is a featureless fluid, and the whole system of corks might be drifting east at 10 miles per hour or west at 20 miles per hour. There would be no easy way to say which individual ant is truly fixed. And no ant could say he was at the center, unless he could measure the position of every other ant and map out the edge of the flotilla of corks.

In reality, astronomers—the analogs of the ants—find no end to the system of galaxies, no edges. We do not even have a fluid medium as an analog of the ocean. All we have is empty space. So it's impossible to say that our galaxy is in a more special, fixed, central or favored position than the Andromeda galaxy or some other galaxy as far away as we can see. We all seem to be floating in an immense ocean of galaxies with no discernible edges or limits.

In the 1930s British astrophysicist Sir Arthur Edding-

Cerro Tololo Interamerican Observatory

Photographs of the two most common types of galaxies. The top photograph is of a typical spiral galaxy. This example is cataloged as NGC 7217 and lies in the direction of the constellation of Pegasus, the winged horse. Its spiral arms are probably wound somewhat tighter than those of our Milky Way. The bottom photograph is of a giant elliptical galaxy. This photograph, which overexposes the billions of stars in the central region, clearly shows the swarm of globular clusters that surround the galaxy like moths around a streetlight. The bright region is roughly 30,000 light-years across. The galaxy is cataloged as M87. It is one of the most luminous of all galaxies and contains an estimated 40 times as much mass as our Milky Way.

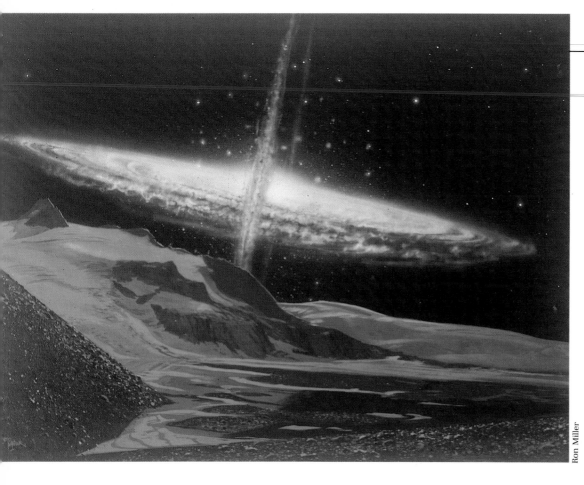

A ringed galaxy fills the sky of an imaginary nearby planet. A few galaxies appear to have a well-defined disk shape, indicating a rotational plane, but also to be surrounded by a nearly perpendicular ring of material. Theorists have suggested that these and other exotically configured galaxies may be products of collisions or mergers of galaxies.

ton gave the name "expanding universe" to this phenomenon, but a better name for the observed fact would be "mutual recession of galaxies." What does the mutual recession tell us about the universe-at-large? Amazingly, some mathematician-astronomers, imagining theoretical pictures of the universe *prior to* the discovery of the mutual recession, had come up with theoretical models that predicted it. At the time, these models were hardly more than mathematicians' amusements, but when the mutual recession of galaxies was discovered in the mid-1920s, astronomers rushed to reexamine the older work.

One example was work by Albert Einstein and others. Einstein's predecessors had shown that, in most models with stationary galaxies, there was a tendency for gravity to pull everything together. The implication was that the universe would eventually collapse with a great, cosmic crash! Since the universe had obviously not crashed, Einstein speculated that there might be an unknown force in nature that caused matter to repel other matter over huge, cosmic distances, thus preventing the final crash. If matter could attract other matter through normal gravity over short distances, why could there not be a weak repulsion at large distances? Under certain mathematical assumptions, the repulsion could be such as to cause galaxies separated by long distances to start moving away from each other. The farther the distance, the stronger the repulsive force and the faster the recession, as has been observed.

But later work led astronomers and theorists to abandon the idea of a repulsive force. New evidence indicated a still more incredible explanation. Sixteen billion years ago, give or take a few billion years, there was a cataclysmic and possibly unique event: a giant explosion that started everything. All matter is spreading out through the universe because it was "launched" outward by this primeval explosion. The primeval event is called *the Big Bang*.

THE BIG BANG:
THE BEGINNING OF EVERYTHING

One line of evidence for a unique primordial event—the Big Bang—is easy to understand. Suppose you made a time-lapse movie of the universe, showing the galaxies receding from each other, and then ran the movie backwards. It would show the galaxies getting closer together. How far back in time would you have to run the film before the galaxies were touching? Estimates come from attempts to measure the rate at which the galaxies are moving away from us, as a function of distance, and these measurements are controversial. An average galaxy a million light-years from us is probably receding at a speed of around 21 kilometers (13 miles) per second. Using such a number, we can calculate how long it has taken for the galaxies to reach their present distances. Estimates of the age of the universe have yielded a figure around 16 billion years for all galaxies, but in view of all the controversy it might be safer to say the figure is between 12 billion and 20 billion years. For convenience, we will refer to the 16-billion-year figure without qualifiers—but remember that the seemingly precise language of science always needs a qualifier if viewed at a fine enough level of discrimination. In science, "16" always means "16 plus or minus such and such uncertainty," whether the uncertainty is 4 or 0.00004.

The moment of maximum density 16 billion years ago, at which all material began expanding, is called *the Big Bang*. It corresponds closely to the Western tradition of a Creation, although researchers have not been able to rule out the possibility that there might have been cycles of expansion and contraction even before the Big Bang—many big bangs, each followed by expansion, galaxy formation, and then ultimately recollapse, making a new big bang and starting a new cycle. This picture would correspond somewhat to Eastern traditions of eternal cycles of the universe. It is not our goal, however, to try to fit nature's facts to some preexisting ideology; rather, we want to find out the facts of what happened and then fit our ideologies to them. Better to let ideologies follow facts than facts, ideologies.

Additional evidence supports the remarkable idea that the galaxies—or, more accurately, the material from which they formed—started rushing apart roughly 16 billion years ago. Physicists, such as the Russian-American researcher and popular writer George Gamow, realized that if the material of the universe was once scrunched together at much higher densities, it must have been extremely hot, for the same reason that a protostar heats up as it collapses. One can do a calculation. If one imagines the gas of the universe at its present mean temperature, and then imagines it having fallen together to re-create its condition at some primordial time (a late frame in our reversed time-lapse movie), one can calculate the temperature and pressure conditions in the past.

Physicists quickly realized that they were in the extraordinary position of being able to talk about events during the first *seconds* of the universe's existence, after the Big Bang! The calculations show that matter would have been extremely hot, completely broken down into

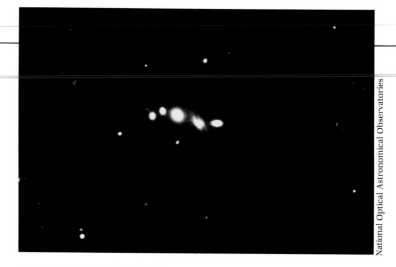

A photograph of a peculiar galaxy with a ring around it. This galaxy looks like a flattened elliptical galaxy of stars, with a ring of dust and stars around it at an inclined angle. The causes of such unusual shapes among a few galaxies are uncertain, but they might result from collisions between galaxies. The painting on page 138 shows how one of these galaxies might look to a nearby observer.

This photograph shows one of many clusters of galaxies—a group cataloged as VV172, forming a string of five in close proximity. Like stars, galaxies apparently formed in clusters, loosely grouped throughout the universe.

subatomic particles and radiation of very short wavelength. Using data on nuclear reactions, physicists showed that the hot, expanding gas—not unlike the inside of a star—was an environment where certain elements could be created by fusion. The calculated relative abundances would explain some features of the observed abundances of various types of elements, although most elements' abundances reflect the later fusion processes inside stars. The agreement with these observations supported the theory that the universe came into being through a big bang.

The clincher that caused virtually universal scientific acceptance of the Big Bang theory was the discovery that the sky is giving off very weak radio radiation of a type predicted by the theory. The explanation of this radiation is complex, but it can be thought of as a view of the flash that accompanied the Big Bang. The idea is as follows. As we look toward more and more distant galaxies, the light from them has taken longer and longer to get here. Light reaching us from a galaxy a million light-years away was emitted a million years ago. Light from a galaxy a billion light-years away was emitted a billion years ago. Looking

between galaxies with our radio telescopes, if we could see light from a gaseous region 16 billion light-years away, it would have been emitted 16 billion years ago, just after the Big Bang.

For a while afterwards, the universe was filled with brilliant light and opaque gas. The reason we don't see this brilliant light, brighter than the surface of the sun, from all over the sky, involves the Doppler shift. The galaxy a million light-years away is receding at 21 kilometers (13 miles) per second and is slightly Doppler-shifted toward the red. A galaxy a billion light-years away is receding at 21,000 kilometers (13,000 miles) per second and its light is moderately red-shifted. The material nearly 16 billion light-years away—the opaque gas that emitted the brilliant visible light—would be receding at a large fraction of the speed of light and the light associated with it would be very strongly red-shifted into the radio region of the spectrum.

Using somewhat similar reasoning, scientists at mid-century predicted that the Big Bang was accompanied by light, whose red-shifted traces might be observed in the radio spectrum. The radiation was discovered in 1965 by

Bell Laboratory researchers and Princeton astrophysicists analyzing microwave noise from all over the sky, detected by the first Telstar communications satellite. It has been called the "three degree radiation" because it has properties of radio radiation that would be seen from material at the low temperature of 3 K ($-454°$ F). The perceived temperature is so low because the radiation has been Doppler-shifted so far toward long wavelengths, hence having the appearance of radiation from a cold body instead of from the original hot fireball. A Nobel Prize was later awarded for the discovery of this radiation, which seemed to confirm the Big Bang description of the origin of the universe.

Working with the Big Bang idea, theorists put together the following history of the universe. Matter and radiation were initially concentrated at very high density. Much of the present mass was then in the form of energy, or radiation, whose equivalent to mass is known through Einstein's famous equation: $E = mc^2$. Within a tiny fraction of a second after the expansion started, the form of the matter-energy mixture—or, more properly, the basic space-time fabric of the universe, as mathematical theorists describe it—went through basic changes and resulted in the kind of space we experience now. For perhaps a thousand years there was more energy per cubic inch in the form of radiation than in the form of matter, but eventually this changed and the universe became like it is now: matter dominates over radiation. Once matter gained the upper hand, the subatomic particles began to aggregate into simple atomic nuclei of hot, ionized gas at temperatures of millions of degrees. The matter in the universe was about three-quarters hydrogen and one-quarter helium. This is why the oldest stars are made of hydrogen

Ron Miller

Before space was black. As presently conceived, the Big Bang could not be witnessed from "outside"—it was a flash of light that filled all of space, and space itself began to expand at the instant of the flash. For the first years, as space expanded and distances in the universe reached dimensions of thousands of light-years, there was only thin, hot gas, bathed and heated by brilliant light. The radiation gradually grew less intense and longer in wavelength, or redder in color. Today we can "see" the Big Bang only as radio wavelengths, and space looks black to the unaided eye. As the radiation intensity faded, the gas became cooler and filaments of gaseous matter began to dominate the universe. The scene viewed here perhaps occurred a few million years after the Big Bang. Eventually, these filaments of dark gas lacing the universe would break up into strings of galaxies—a pattern still detectable in the arrangement of galaxies today.

and helium, with very few heavy elements; the heavy elements had to await formation in the pressure cookers inside stars. The subsequent explosion of these stars threw heavy elements into space.

After a few million years, the universe became even more recognizable. The gas temperature had dropped to a thousand degrees or so, and the electrons were joining the nuclei to become neutral atoms instead of ions. The stage was now set for what we have already described: by the end of the first billion years, the gas began to collapse to form galaxies. Our particular galaxy arose after about 3 or 4 billion years, and our particular star and planet formed in that galaxy after about 11 billion years. Another 4 billion years was required to make large animals. We humans came along in only the last couple hundredths of one percent of the history of the universe.

AN ALTERNATE THEORY

Some people in our society are uncomfortable with this sort of picture. They prefer to accept the more literary picture of ancient traditions presented in the Bible, the Koran or the Bhagavad-Gita. They prefer to say simply that the universe was created by a creative essence, God, and that's the end of it. Some of these fundamentalists (be they Christian, Moslem, Hindu, etc.) become incensed at the language above because they insist on reading literally the words of their chosen tradition. If their chosen scripture—usually the one "chosen" by the culture into which they were born—says that creation took seven days to lead to humanity, then they say that this language is talking about not seven metaphorical stages but exactly seven literal days. Therefore, they say, all talk of billion-year periods of galaxy formation is ideologically incorrect. If their preacher interprets the scriptures to mean that the earth formed in 4000 B.C. (as a famous bishop did in the 1600s, yielding an idea that persists among Christian fundamentalists even today), then all the talk of a 4.5 billion-year-old Earth or million-year-old Homo sapiens or even the evolution of species must be wrong. And all the scientific data must be wrong and should therefore be ignored. Fundamentalists have initiated lawsuits to have this non-astronomical approach taught in our public schools. Fortu-

nately, we live in a part of the world where all are free to pursue their own beliefs, as long as they don't impinge on the rights of others to free discussion and information. We hope that this right will keep our government from decreeing that some particular religious minority's view *must* be taught in public schools, as has happened in totalitarian countries.

To each new theory or political challenge, science has an answer: let everyone put his or her evidence on the table. The "winner" is the person who puts the best and the most convincing evidence on the table, not the person who wields the most political clout or makes the most clever debating points. The answer of scientists is that we should freely look at the evidence for each point of view. If fundamentalists (Moslem, Christian, Marxist, or whatever) choose the philosophic position that only their ancient texts are relevant, that nature's evidence should be ignored, the scientists' response is: an exciting life is a dialogue with nature, a learning from the world around us. Nature or God (whichever term you like; both relate to a single, similar concept) has strewn the world around us with clues. Fundamentalists claim they are false clues; maybe all those galaxies rushing away in the sky, and those isotopes that yield 4.5-billion-year-old dates in meteorites, were put there by a capricious god to fool us. Scientists prefer to study and act on the clues.

Scientific clues tell a story, and it's an interesting story; we are interested in reading the stories written in nature by the gods! They are stories that lead us down a fascinating road, and we are fascinated and alive if we follow the road. It's like a treasure hunt: go to the foot of the tree; you will find a note that tells you to look behind the loose brick in the wall; and so on. Point your telescope to the sky; you find the galaxies are rushing away; calculate the time it takes and you find an age for a creation event; calculate the

Clouds and clouds. Cumulus towers on a giant, hydrogen-rich planet are silhouetted against clouds of a different sort—the vast star-clouds of the bright central regions of a spiral galaxy. The planet is located on the outskirts of a cluster in a small satellite galaxy, some of whose stars are seen in the intermediate distance.

William K. Hartmann

Some building blocks of the universe. A double star consisting of a red giant and a white dwarf, with its surrounding, flattened dust system, is superimposed on a massive spiral galaxy. The foreground disk contains dust blown out of the red giant. Much of it may eventually be dispersed, forming interstellar dust. The distant galaxy is made up of a hundred billion star systems, stars with a vast range of sizes, masses and colors. Older, redder stars are concentrated around the galaxy's nucleus, while the outer spiral arms are dominated by short-lived, massive bluish stars clumped in regions of fresh star formation. Unnumbered remote galaxies populate the endless reaches of background space.

The universe of galaxies. Recent observations have shown not only that galaxies are clumped in clusters, but also that the clusters are strung out in a filamentary pattern. Astronomers believe that this wispy filamentary pattern stems from the structure of gas in the initial universe, as the matter formed shortly after the Big Bang.

Ron Miller

conditions in the creation event and you predict three-degree radiation; point your radio telescope at the sky... and *voilà!* you find the three-degree radiation! This treasure hunt allows us to make predictions, and sometimes we're delightfully surprised when the prediction helps us discover something new. The predictions of science have had more success than the predictions of fundamentalists (which have included numerous claims of an imminent end of the world).

The cosmic treasure hunt of science helps us learn how the universe acts, and this in turn helps us learn how to build spaceships, bridges, stereos and indoor plumbing. Reliance solely on ancient traditions and refusal to look nature in the face does not teach us these things.

This is not to say that science teaches us how to live. Science/technology also produced MIRVed warheads, which are not healthy for living things. But at its best, science can help us learn how to live in the universe that the Big Bang created. Science does not teach us the whole

story of how to live, but it does at least give us guidelines for testing our belief systems against the real world. I assert that it's better to allow a vigorous, growing ideology to emerge from our knowledge of the universe—that is, to learn how to act consistently with how matter, energy and your body work—than to try to force the universe to fit any preconceived ideology.

THE "SHAPE" OF SPACE

Many people assume from descriptions of the Big Bang that if someone could have been there at the time, one could have stood aside at a "safe" distance and witnessed the titanic explosion that started the universe. This was impossible, of course, because the explosion contained *all* of the universe; you couldn't stand aside. But the full answer is a bit more subtle than that. We should not draw a picture of a fireball expanding out into empty space; the

Ron Miller

In their ponderous wheelings past each other in clusters, galaxies sometimes collide. Computer studies have revealed that tidal forces during the encounters pull out streamers of gas, dust and stars. These streamers are deflected in wild arcs reaching far out of the system. Astronomers have whimsically dubbed these "rat-tail" galaxies.

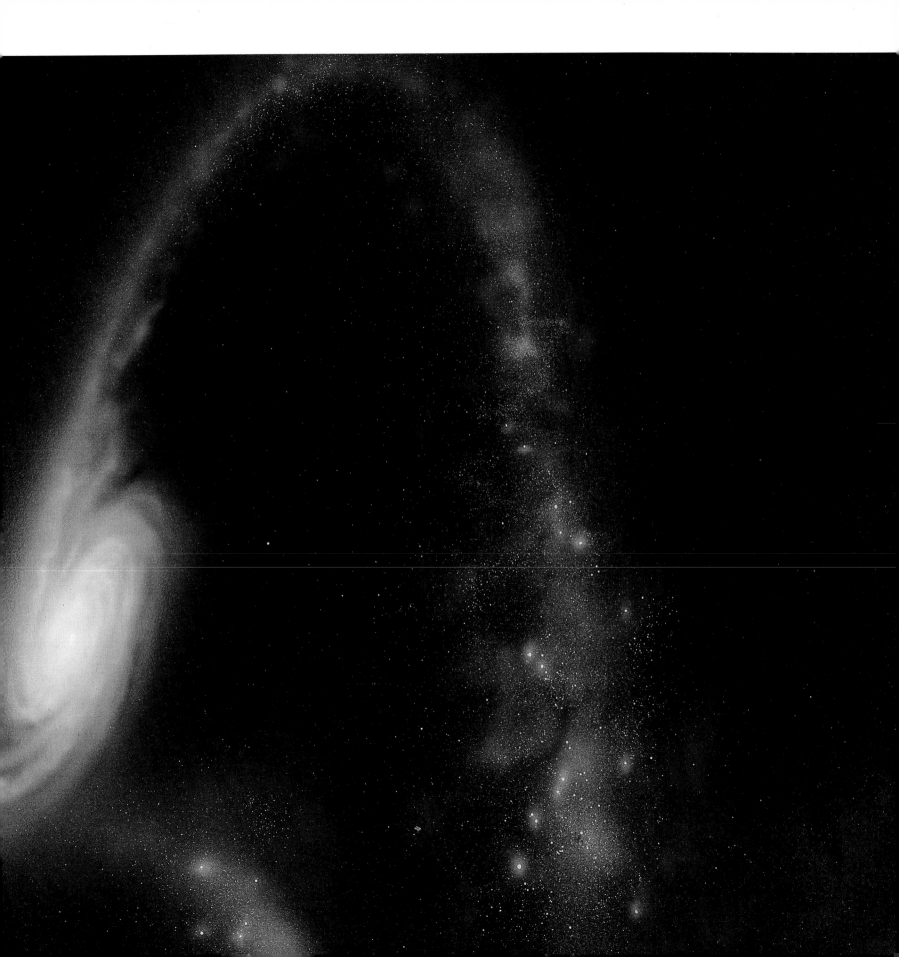

conception of cosmologists, those theorists who picture the structure and beginning of the universe, is that the fireball filled all of space, and space itself began to expand.

To visualize this, we need an analogy. We can imagine three-dimensional space as a two-dimensional surface. For example, let us return to our earlier image of the galaxies as corks floating in an ocean, with an ant-observer on each. The corks are allowed only on the surface of the ocean, and the two dimensions of latitude and longitude can specify the position of each. The ocean surface corresponds to their conception of space, since the ants can move in any direction on that surface. They look out and sense that the cork-galaxies go on forever, out to the flat horizon, as far as they can see. Perhaps it truly is flat, as

A photograph showing part of the huge Virgo Cluster of galaxies. This cluster, in the direction of the constellation Virgo, is believed to contain some 2,500 galaxies and lies roughly 60 million light-years from our Milky Way!

National Optical Astronomical Observatories

flat as an infinitely big chessboard. But eventually some ant-Euclids and ant-Magellans may discover that the surface—their "space"—is curved. They could find this out in several ways. If they sailed off in one direction, in principle they would come back to where they started. Or, if they had ant-telescopes to count the number of corks within a given distance—one mile, ten miles, a thousand miles, ten thousand miles—they would find out that the number of corks per square mile did not keep increasing with the square of the distance as would have occurred had their space been flat. Indeed, in that case there would be no more *new* cork-galaxies to count at a distance greater than half the Earth's circumference; their universe would be finite and have a finite number of galaxies in it.

So far the analogy is quite apt. In fact, cosmologists debate whether our three-dimensional space is flat or curved in the sense described above. It is actually open to testing: astronomers attempt to count galaxies within various distances to see if the relationship corresponds to flatness or some curvature. So far the counts indicate a close approximation to flatness, but there is a possibility of slight curvature, and studies continue from year to year. The "radius of curvature," of course, could be so enormous that space is nearly flat and looks flat to us on our local scale, just as the calm ocean looks flat to the ant-observers.

One day the ants make a stunning discovery. They find that the Earth itself is expanding. This is somewhat analogous to many cosmologists' concept of the universe. Thus, instead of the corks or galaxies simply moving apart "through" space, there is at the same time a change in curvature of space itself. The ants now imagine the past and realize that their whole Earth-universe was smaller in the past. It had less area—or, as we would say in our universe, volume. When their Earth-universe was only a mile across, all the corks were much closer; when it was the size of a ping-pong ball, all of the cork-matter was mashed together and disassembled in a blaze of light. Still, an imaginary ant-observer of that time could only have existed on the surface of the ping-pong ball and hence would have been immersed *in* the blaze of light—the ants' Big Bang. To "stand aside" and "observe" the ping-pong ball from a distance, the ant would have had to enter another dimension, outside the familiar two-dimensional, latitude-longitude universe. But no ant could travel in such a di-

Ron Miller

From the surface of a planet of a star torn out of one galaxy during its collision with another, we look back at the parent galaxy being transformed into a doughnut-shaped ring. The gravitational interactions of two galactic systems as they pass through each other cause major distortions, which sometimes create odd shapes such as this ring. The tail end of the galaxy that did the damage can be seen in the far upper-right corner of the picture.

mension. Similarly, at least from the point of view of many theorists, we cannot speak of looking at the primordial Big Bang fireball from a distance because we and everything else were immersed in it.

A CYCLIC UNIVERSE?

And what of the future? The galaxies are now rushing apart. Will they continue to do so? Will the universe thin out into an even sparser nothingness than it is now? Or will the galaxies slow down and start to fall back together into another vast fireball? Is the story of the universe just one giant cycle of fire, or a succession of many cycles?

Early in the book we spoke of a first principle: gravity rules the universe. This simple principle we used successfully to explain why the primordial gas broke into clumps that formed galaxies and why the gas inside galaxies breaks into clumps to form stars. But does gravity rule the *history* of the universe? Will gravity win and reverse the mutual recession of the galaxies?

Astrophysicists attempt to answer these questions by studying the total density of matter in the universe. If the density is too low, there is not enough mutual gravitation to reverse the flight of the galaxies, they can expand into space forever, until all the fuel that can be "burned" in all the stars *is* consumed in nuclear reactions and the stars are reduced to cooling ashes: white dwarfs, neutron stars, black holes, brown dwarfs, cold planets and various subatomic particles. But if the density is higher than a certain critical value, then there is enough mutual gravitation to

slow the galaxies and reverse their flight; they would eventually start to fall back. In this case, billions of years into the future, there would occur the reverse of the Big Bang: the big crunch. And what then? All galaxies, stars, planets, life, would be extinguished in a flash.

Some theorists, however, have speculated that the universe might be cyclic. Just as old humans and animals die but give birth to a new generation, just as stars die but spew out material for a new generation of stars, the universe itself may be cyclic on a vaster time scale than we had imagined. Perhaps this universe of stars consumes itself but collapses to initiate a new Big Bang. Perhaps the future holds new universes, new cycles of stars and planets, new civilizations forever disconnected from ours by vast cosmic events of destruction and rebirth.

Present-day measures of the *directly observable* material in the universe reveal much less than the critical density. If that were all the material there is, the galaxies should recede forever. But astronomers are confident from observing the motions of galaxies that the galaxies contain much mass that cannot be seen. No one is sure in what form this material exists, whether it is asteroid-size bodies or subatomic particles. Some astronomers, as a result of analysis of the unseen material, think the total density of the universe is quite close to the critical value, meaning that the future is still unknown! Through the orbiting space telescopes of the next decade, astronomers hope to have a clearer view of the most distant parts of the universe, where they can measure the total density of matter and gain a better understanding of the possible fate of the material system in which we live.

A WINDOW INTO THE PAST

We have seen that the Milky Way galaxy has an extraordinary central region known as the nucleus, where evidence suggests highly energetic events. Other galaxies also have brilliant nuclei—small, central regions only tens of light-years in diameter that emit much of the light of the galaxy. Indeed, some of the most distant galaxies can

be seen only by the light of their nuclei because the outer regions are too faint to be visible.

In the 1950s and '60s astronomers discovered that there is a hierarchy of galactic nuclei, from the relatively common, "normal" ones like ours to vastly more energetic ones. This discovery came about partly through radio astronomy. As we saw earlier, amateur radio buff Grote

A sequence of galaxies, possibly marking different evolutionary stages or intensity levels in the hierarchy of active galaxies. Most normal, or inactive, galaxies have a bright central nucleus (at top). In Seyfert galaxies (second from top) these nuclei are more luminous and variable. Some exploding galaxies (middle picture) show splatters of gas ejected from the nucleus, along with strong radio radiation. In some active galaxies (second from bottom, above), "vertical" jets shoot out of the galactic plane, accelerated by poorly understood processes, probably in disks of gas around black holes in the galactic center. Most brilliant of all are quasars (at bottom), which are probably extremely luminous or exploding nuclei. Some show jets and some may have spiral arms, though their forms are poorly known because they are so distant. The source of their luminosity is unknown, but may involve collisions between stars or gas clouds having enormous black holes and millions of solar masses of material.

A jet in the heart of a giant elliptical galaxy. This photograph shows the central core of the galaxy cataloged as M87. The jet of luminous gas is also a powerful emitter of X-rays. The process that focuses and accelerates gas in galactic jets is not well understood.

Reber was the first to map radio radiation coming from our galaxy and to discover that the maximum radiation came from the center. Soon, astronomers found that some of the strongest distinct sources were other galaxies. Only certain galaxies were in the new category of so-called *radio galaxies*. In many of these, the radio astronomers reported particular types of radio emission characteristic of hot gas interacting with strong magnetic fields. Quickly, other astronomers used traditional telescopes to photograph these odd, radio-emitting galaxies and to find out how they were different from ordinary galaxies. What process was heating and exciting their gas, causing emission of the particular types of radio waves?

Photographs revealed that one class of radio galaxies were *colliding galaxies*. Galaxies have random motions, relative to each other, in addition to their general mutual recession. By chance, two galaxies might be moving through each other. The individual stars of galaxies are too far apart to be likely to collide—just as two swarms of gnats might move through each other without individual gnats colliding—but the gas clouds between the stars col-

lided and the resulting excitation of the gas caused the radio radiation. Colliding galaxies, however, turned out to be only a subset of the whole sequence of radio-emitting galaxies.

ACTIVE GALAXIES

Photographs revealed that a larger, more important class of radio galaxies displayed strange, explosive features around their nuclei. Some of these had unusually bright nuclei. Others showed ragged splatters of glowing hydrogen gas, as if the gas had been explosively ejected from the central region of the galaxy. Still others showed jets: long streamers of hot, glowing gas, shooting directly out of the center in two opposite directions. All galaxies of these types, with disturbed, extra-bright centers, are called *active galaxies*. At least 5 percent of all galaxies are active. Although the nucleus of our galaxy emits powerful radiation, it is not as powerful as the nuclei of active galaxies. Thus our galaxy is not active but "just" a normal galaxy.

Astronomers now believe, on the basis of these galaxies' statistics, that the kinds of events we described in the center of our galaxy occur in most galaxies and occasionally lead to titanic outbursts as well as heating and ejection of gas. Gigantic black holes may exist in the nuclei of many or most galaxies. Gas becomes heated and compressed as it spirals inward to the region of the nucleus,

On a planet in a jetting galaxy. This planet's star orbits around the galaxy in a somewhat inclined path, taking it to a higher position above the galactic plane than we are above the Milky Way. This affords a view "down" into the central region from a higher angle than we have above the intervening dust in the Milky Way, thus revealing the central bulge of reddish stars and a glimpse of the central region. In the Milky Way, this view is blocked by dust clouds. As twilight fades beyond a high mountainside's volcanic cinder cone, the milky band of this alien galaxy emerges in the night sky. The galaxy's luminous jets stretch in opposite directions from the galactic center in ghostly searchlight beams across the sky.

William K. Hartmann

forming a hot, glowing accretion disk. Much of the gas may eventually fall onto a black hole or giant star at the center of the disk. Other gas, perhaps caught up in magnetic fields, squirts out from the "north" and "south" poles of the accretion disk, forming high-speed, narrow jets moving in opposite directions. The intense energies involved are illustrated by the fact that some of these jets probably move with an appreciable fraction of the speed of light!

An example of the evidence for supermassive objects in the nuclei of galaxies came in 1983 when European astronomers, observing the velocities of gas clouds orbiting the center of an active galaxy by using Doppler shifts, happened to observe a flare-up of the nuclear region over a period of a few weeks. Combining the orbital velocities they had measured with the rate of expansion of the flare-up, they were able to measure orbital speeds within a tenth of a light-year from the nucleus—a region no bigger than that occupied by the cloud of comets around our solar system. In other words, they could measure properties of a region around the nucleus that was very "tiny" by galactic standards. Since the orbital speed depends on the mass of the central body, they could calculate the mass within this solar-system-size nuclear region. They came up with a fantastic 100 million solar masses—the mass of 100 million stars crammed into a region that would easily fit between the sun and Alpha Centauri! It is these kinds of numbers that force astronomers to talk of extraordinary conditions in galactic nuclei.

The least energetic of the active galaxies is a type with an unusually bright nucleus; these galaxies are called Seyfert (pronounced *see-fert*) galaxies, after their discoverer. They have brightness fluctuations and hot gas ejected at speeds of thousands of kilometers per second. Perhaps we are seeing explosions resulting from the fall-in of individual stars onto the nucleus' accretion disk. The galaxies with explosive splatters and jets of gas are still more energetic.

QUASARS

At the other end of the energy scale are the most energetic galactic nuclei. These probably grade into the famous objects called *quasars*. I say "probably" because while most

An exploding galaxy, seen from an imaginary planet system in a nearby satellite galaxy. Extremely energetic explosions have occurred in the center of some galaxies. These are revealed by the galaxy's extraordinary luminosity and by splatterings of hot gas, especially hydrogen, emanating from the galactic center and glowing with a characteristic red color.

Ron Miller

Ron Miller

astronomers now think quasars are intensely active nuclei of very distant galaxies, a few believe that they may be new objects of an unknown type, perhaps satellites of closer galaxies. We will adopt the more conventional view, that quasars are ultraluminous exploding nuclei of very remote galaxies.

The name "quasar" (an acronym for "*quasi-stellar radio source*") refers to the fact that radio astronomers, while mapping radio galaxies, cataloged a number of objects that showed up in photos only as starlike dots. One indication that they might be very remote galaxies came from the fact that they have enormous Doppler red shifts, corresponding to movement away from us at as much as 90 percent of the speed of light or more. Remote galaxies are the only kinds of objects known to have such large red shifts. Further studies revealed hints of spiral arm structure or star clouds around some of the fuzzy, dotlike quasars. This supported the view that they are extremely distant galaxies.

An extreme example of a quasar is one of the most luminous ever observed, cataloged as S5 0014 + 81. This object, which looks like a faint dot in our telescopes, has been estimated to be about 10 billion light-years away on the basis of the relation between red shift and distance among galaxies. According to a 1983 study, its total output of light and radio radiation, assuming this distance, amounts to around 100,000 times the energy output of our galaxy!

The probable "close-up" appearance of a quasar. Although there is controversy about their exact nature, quasars appear to be extremely brilliant exploding centers of galaxies. Some shine with tens of thousands of times the light of our whole galaxy! Seen from "nearby" space somewhat outside its galactic edges, a quasar galaxy would be a brilliant beacon, with the surrounding spiral arms or other features barely visible in the glare of the brilliant nucleus.

Many of the most distant objects we can see are quasars, because only such bright objects can be seen from billions of light-years away. Astronomers have concluded that a higher fraction of the most distant galaxies are active galaxies and quasars than among the local galaxies.

Now, consider a quasar 10 billion light-years away. The time for the light to have reached us from such an object is 10 billion years! We are seeing this object not as it is today, but as it was when the light started its journey, 10 billion years ago. This is an appreciable fraction of the way back to the Big Bang, some 16 billion years ago. It means that when we look at quasars, we're seeing galaxies as they were a few billion years after the Big Bang.

From this reasoning, many astronomers believe that quasars represent an early stage of galaxy formation. Perhaps the central regions of galaxies are turbulent and unstable as the galaxies take shape. Due to drag forces, gas and dust orbiting in the inner part of the galaxy spiral in to the center. Initially, this may lead to a concentration of material and the formation of gigantic stars, perhaps with 100 or 1,000 times the mass of the sun. As we have seen, however, these would be unstable and would explode as supernovae within a few million years. The explosions, in turn, would disrupt the gas. But, inexorably, the galaxy keeps feeding more gas into the central regions. Thousand-solar-mass star corpses might grow into 10-thousand-solar-mass black holes. Perhaps these collide, merge and form million-solar-mass black holes as more gas is dumped upon them.

The study of active galactic nuclei, galactic explosions, gas jets and quasars is one of the most exciting areas of current astronomy. Astronomers try to comprehend some of the highest-energy events known in the universe. These objects are interesting in their own right: they show us extraordinary events and may help us understand new techniques of energy production for our own use on Earth, as we learn how the magnetic fields bottle up hot gas and make it flow in certain directions. But beyond that, they give us a window into the past. As we look at remote quasars with today's biggest mountaintop telescopes and tomorrow's giant orbiting space telescopes, we'll be seeing the earliest days of the universe when gas blown out of the Big Bang was collapsing helter-skelter into ponderous disks and ellipsoids of stars—the first, and most violent, galaxies.

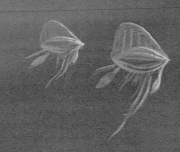

THE POSSIBILITIES OF

THE UNIVERSE

Some scientists, including Carl Sagan, have suggested that jellyfish-like floating organisms could have evolved in the dense, organic-molecule-rich, lower atmospheres of giant planets such as Jupiter. Here, in the lower left-hand portion of the picture, a pair of such eel-like "floaters" swim (or slither) in the thick atmosphere of a gas-giant planet of some other star. [Preceding page]

Ron Miller

ALIEN LIFE AMONG THE STARS?

Earlier we spoke about the Copernican revolution displacing Earth from the center of the universe, the Darwinian revolution displacing humanity from a unique position outside the animal kingdom, the Shapley-Hubble revolution displacing us from the center of a supposedly unique galaxy, and a possible revolution coming in the next decade—the confirmation of extrasolar planets that would displace us from living in a uniquely created system of worlds.

There is a fifth potential revolution that may or may not occur: the discovery of life beyond Earth. This would displace us from the biological center of the universe. A subset and perhaps an inevitable consequence of this revolution would be the discovery of life that is more intelligent, or more technically advanced, than ours. This would be a stunning, though not necessarily devastating discovery—rather like your grade-school discovery that the kid down the block gets better grades or plays the piano better than you do.

Yet perhaps, in the long run, it could be devastating. After Europeans first contacted Hawaii in 1778, the native Hawaiian population dropped to half its former level within fifty years because of diseases that were introduced. Some 95 percent of the native population of Guam was similarly wiped out within a century of European contact. The Plague, caused by bacteria introduced into Europe in the 1300s, killed about a quarter of the European population. Similarly, we might worry about viral disease transmission to terrestrial lifeforms, if we should come into contact with aliens or alien organic material. For similar reasons, Apollo 11 astronauts were quarantined when they returned from the moon, although the moon dust turned out to be sterile.

Even without the threat of disease, there is the issue of the kind of cultural competition that arose between the native Americans and the European settlers—to the detriment of the Indians. Not only did military conflict occur, but some of the European invaders made explicit efforts to

Pamela Lee

Subtle evidence of an alien civilization. An orbital view of a planet at dusk reveals the lights of a settlement pattern of highly evolved creatures on the surface. The planet is in the system of a sunlike star that has just set, casting a sunset-colored band along the horizon. A companion star, or substellar object, in the system is the distant, red-glowing object at the upper left. A moonlike satellite of the planet is illuminated by both light sources.

demolish the native culture as well. Spanish priests, for example, tore down many meso-American temples and built Christian churches atop the ruins in order to destroy paganism; they also burned virtually all the priceless Mayan books, which recorded astronomical observations and other records. Thus anthropologist D. K. Stern remarked in 1975, "It is likely that the meeting of two alien civilizations will lead to the subordination of one by the other." Of course, one can always imagine that a civilization more enlightened than ours would recognize the dangers of contact and treat us gently, or perhaps avoid contact altogether. Since we have no concrete data—not a single example of extraterrestrial life—the possibilities for speculation are endless. They have spawned endless science-fiction tales.

Viewed in this light, the most prudent approach to alien intelligence—and perhaps the only step we can take now, anyway—might seem to be to search for signs of alien life without revealing ourselves. In this way, we could assess the possibilities before initiating contact.

FALSE START: UFOS

But perhaps we do not have the luxury of choice. The revolution associated with "first contact" could come about in two ways. We could discover extraterrestrial life—or it could discover us. The second possibility brings up the UFO (Unidentified Flying Object) phenomenon, which many have claimed is already evidence of the existence of intelligent aliens! I have tried to approach this controversy with an open mind. In the late 1960s I served as an investigator with the U.S. Air Force-sponsored, comprehensive study of the phenomenon at the University of Colorado. I joined in the hope that we could find solid

An alien planet has produced tree-fern-like vegetation, seen here under an evening sky dominated by a nearby Mars-like giant moon and by a more distant smaller moon. At the lower right another planet in the system makes a striking evening star. We don't know the nature of the phosphorescent organisms deep in the underbrush.

evidence of something mysterious. I found that there was more tractable science in the *sociology* of the phenomenon than in the *physical evidence*. For instance, we might talk with a witness who insisted he took two photos, a few seconds apart, of a UFO swooping overhead, although our photoanalysis had already shown from shadow changes in the pictures that they were taken an hour apart. The totality of such evidence seemed to establish a fabrication, and with such knowledge in the back of our minds it was strange to conduct a pleasant, seemingly serious conversation with the "witness," who insisted he had really seen a spaceship.

The experience of being a UFO investigator also affected my view of the media in our society. It was frustrating to read a front-page story on some mysterious sighting in some state and then, when the event was conclusively identified two weeks later as a sighting of a weather balloon or a satellite reentry, to notice that the same newspaper carried a tiny item on page 12 recounting the explanation. It was on page 12 because, from the point of view of the editor trained in the traditions of American journalism, this.was "yesterday's news." The public, meanwhile, usually fails to notice the later page-12 item; the public ends up ill-informed as a result of this common journalistic practice. After years of exposure to the "front-page mysteries," which are never adequately followed up by the press, the public accumulates the impression that there is a vast body of unexplained sightings of celestial phenomena. Even today, to boost their sales, magazine and tabloid editors frequently run a "flying saucer photo," often on the front cover, which in fact is several years old, has been investigated and has been explained. Sometimes these are natural phenomena (a famous "UFO" photo of a lenticular cloud is an example), and sometimes—worse yet—they have been established in published literature as fakes. While a few of the UFO reports do suggest some credible observations of unusual atmospheric events, possibly electrical in some cases, there seems to be, in my opinion, no credible *physical* evidence for interstellar spaceships.

Given the number of military sensing networks and amateurs abroad with cameras at all hours of the day and night, one might expect that if spaceships were zooming around with the frequency sometimes imagined by UFO

enthusiasts, we would have some evidence by now. For example, when a fireball streaked across the western states in 1972, hundreds of eyewitnesses gave consistent reports, eleven different still photos and some movie frames were published in an astronomy magazine, and a classified Air Force satellite detected the object from above. In the absence of such evidence for alien spaceships, we have to conclude that while a visit from an interstellar spaceship is plausible, and could happen at any time, we have no evidence of its ever having happened.

WHAT ARE THE ODDS OF FINDING ALIENS?

So we come back to the first possibility: that *we* might eventually discover *them* before *they* (if they exist) make themselves known to us. What are the chances of this? Usually, we restrict the discussion to life as we know it, based on the ability of carbon atoms to link into long chains, creating complex molecules that, by their ability to add other molecules, divide and replicate, form the basis for living organisms. As for life as we *don't* know it, such as sentient interstellar clouds or intelligent rocks: maybe so, but we don't have much evidence on which to base a discussion. If we stick to carbon-based life, we can at least assess by experiments and astronomical observations whether the right kinds of molecules, and the required environmental conditions, have arisen elsewhere.

There is a famous approach to this problem, popularized by radio astronomer Frank Drake. The so-called Drake equation reduces the amount of wild speculation by systematically assessing the probabilities of each link in a chain of necessary conditions for life to have arisen outside our solar system. These conditions include the existence of planets, the necessity for a habitable climate, the necessity that such a climate remain reasonably stable long enough for life to evolve, and so on. The Drake equation does not reveal new information that we did not already know, but it does provide a framework for utilizing known information in a systematic way.

Following a similar approach, let us ask how many planets in our own galaxy might plausibly harbor intelligent life. We start out with the astronomical measure-

Many current theories of the origin of life invoke the concentration of organic molecules in primordial oceans. Formation of such molecules is aided by energy input from lightning bursts. Here we imagine the next phase—organic slimes floating on the watery surface of a young, Earth-like planet somewhere else in our galaxy. In another two billion years there may be fish swimming in these seas.

An alien sunset from the rim of a flooded crater. Plant life has gained a root-hold on this fertile world, where volcanism and ocean erosion have combined to create flooded volcanic craters that harbor microenvironments rich in organic materials.

William K. Hartmann

ment that our galaxy contains about 200 billion (200,000,000,000) stars. To be conservative, let us follow our earlier discussion and assume that double- and multiple-star systems have no planets and that only stars similar to the sun (say, with one-tenth solar mass to two solar masses) evolve planetary systems. Perhaps in the rest of the cases the planet-spawning dust blows away before planets have a chance to form. This would still mean that roughly 5 percent of all the stars in our galaxy have planets. Just to be extra conservative, let's call it one percent. We would still have 2 billion systems with planets in our galaxy alone!

To generate life, we probably need liquid water, since biological and laboratory evidence suggests that a liquid water medium was helpful in allowing organic materials to form and aggregate. In our system, one out of ten planets has a water-bearing, clement environment. It would be two out of ten if we count Mars, since Mars apparently did at one time have liquid water running on its surface. Again

being conservative, let's call it 10 percent, leaving perhaps 200 million systems with planets where life could start.

All laboratory experiments to date suggest that organic molecules and complicated organic sludges form rapidly under a wide variety of conditions as long as the basic ingredients are present. These include water, carbon compounds, certain gases like ammonia and methane, and an energy source such as lightning or strong sunlight. Even meteorites contain amino acids, the building blocks of protein. So the general thinking among biologists is that life will begin whenever it is given an environment where it *can* begin. But let's be conservative and say it starts only half the time. That leaves 100 million systems with planets on which primitive living organisms formed.

The next key to the argument is the stability of planetary climates. If planetary climates are not stable, climate disasters might be a crucial factor in further reducing the number of planets on which life has survived. The climate on Mars seems to have changed dramatically, and even on

Pamela Lee

Life on this planet was extinguished when a close encounter with another large interplanetary body made the planet's orbit more elliptical and dropped the mean temperature to below freezing during a long winter at great distance from the central star. Or are some living creatures, adapted to hibernation during this long winter, waiting for the brief spring and summer when the planet swings close to its star?

Pamela Lee

A perfect fossil, preserved in ice. Judging from what we've learned about Mars and Earth, disastrous climate changes may have occurred quite suddenly on some planets. Hence, an alien creature has been preserved in ice as a planet-wide freeze sets in after some climatic catastrophe. In the ice we see the reflection of a double star overhead.

Earth occasional asteroid impacts probably wiped out many species, so let's suppose that only a tenth of all planets have a stable enough climate to last long enough for intelligence to evolve. So we have 10 million systems with planets harboring evolved life. If half of them are older than our system, we might imagine that half have life further evolved than ours. That leaves 5 million star systems harboring intelligent life, or at least life with a longer history of biological evolution than ours. These figures are only the roughest of estimates, based on our present scientific knowledge, but you can see that it's hard to convince yourself that there are any fewer than hundreds of thousands of planets with highly evolved lifeforms in our own galaxy!

How far away would the nearest aliens' home planet be, assuming that our figures are right? If 5 million stars out of the 200 billion in our galaxy harbor intelligent aliens, then one star out of every 40,000 might be expected to have planets with more evolved life than Earth has. Count that many stars in the sky, and you might have looked at one with an alien civilization! If these figures are right, we would have to fly a spaceship to about 40,000 stars before getting a high probability of finding somebody at home that we could talk to. To explore a volume of 40,000 stars, we would have to fly to a distance roughly 150 light-years from Earth. To put it another way, if we believe these figures, there should be intelligent life within about 150 light-years. The number of stars increases rapidly as we increase the distance, so to be on the safe side we might state, as a bottom line in this argument, that there are likely to be highly evolved species within a few hundred light-years of Earth.

Figures such as these have been widely used in arguments about seeking extrasolar lifeforms. They are fairly convincing. There *could* be evolved life within our region of the galaxy. However, many caveats can be added to the argument. For example, I have stressed in other writings that the "evolutionary time-slot" for finding someone we can recognize as a kindred spirit, who might be interested in conversing with us, is very narrow. This makes the figures more pessimistic. Let me explain. Suppose we insist on seeking intelligent, technological creatures who have a civilization somehow comparable to ours and who have evolved in at least a roughly similar fashion. How should

Life on an alien satellite. This world is a satellite of a Jupiter-like gas giant with a ring system, seen here looming in the sky at sunset. The giant planet and its inhabited moon orbit in turn around a sunlike star that has just set. Landforms on this oceanic, geologically quiet world are principally the rims of impact craters; one partly flooded crater-island can be seen on the distant horizon. The most prominent lifeform seems to be a green plantlike organism that is mobile. Does it derive its energy directly from processing minerals in the surface soils?

we define "somehow comparable"? Looking back at the fossil record on Earth, we are impressed by the amazing fact that Homo sapiens, as a species, is not more than a few hundred thousand years old and that recognizable civilization is only a few thousand years old. These numbers are tiny fractions, less than 0.01 percent, of the 4.5-billion-year history of our planet. "Somehow comparable" might thus mean comparable to our state within the last few thousand years and for some time into the future.

The next question is how long Homo sapiens or recognizable civilization might last into the future of our planet. Suppose the planet lasts as long as the sun is expected to—another 4.5 billion years or so. Taking this 9-billion-year period as a representative "active" lifetime of a viable planet around a warm star, we are trying to ask during what fraction of that time a visitor might find an intelligent species or recognizable civilization. There are two quite different schools of thought, both citing the experience of humanity. One is that, having achieved intelligence and civilization of a sort, we will somehow stabilize, escape the threat that war might destroy us, and adopt social and reproductive policies that effectively end natural selection in our species, thus halting or retarding evolutionary change. In this view, Homo sapiens would last much of the rest of the age of our planet, or perhaps indefinitely by means of interstellar migration. We would expect, therefore, that if we visited a sample of habitable planets, about half of them would be too immature in terms of evolutionary sequence for us to find intelligence, but on the other half of them we would find permanent, stable civilizations that might be interested in talking to us. This conception is consistent with the figures we discussed above and would lead to the conclusion that there should be a civilization within a few hundred light-years. In our whole galaxy, there might be perhaps 5 million civilizations!

I find this conception too ethnocentric. It assumes we are the pinnacle of evolution and that we will retain not only our biological characteristics but also our social characteristics, indefinitely. Consider the other extreme, which is pessimistic. We have built a technological civilization that spends a large fraction of its global annual budget on building and maintaining weapons of death; if the worst ones are used, we will destroy civilization and render Earth less habitable. Keeping in mind that we have already consumed most of the accessible coal, oil, iron ore and other resources of Earth, we see that if we slip into a New Dark Age, either through war or resource exhaustion, it will be very hard to support a second industrial revolution and the rise of a new technological civilization. Therefore, we conclude that technical civilization on our planet might end within a century, in which case it would have lasted only, say, a few thousand years out of the 4.5-billion-year history of the planet, or 0.0001 percent of planetary history. According to this pessimistic view, our last multiplication in the calculation four paragraphs above should be by 0.0001 percent, not 50 percent. Therefore we would conclude that because civilizations last such a short time, there are not 5 million but only an average of about 10 civilizations in our galaxy *at any one moment*. The nearest one might be ten thousand light-years away. Old ones might die out and new ones spring up without any one contacting another.

The truth may lie somewhere between these optimistic and pessimistic extremes. The main point is that to believe that we are the only civilization in the galaxy, or one of only a few, seems to require that we believe that civilizations die out soon after they arise, due to one type of disaster or another.

It is hard to believe that all civilizations die out after only a few thousand years. Even if nuclear war should come to ours, there may be some survivors (villages of Eskimos? Andean Indians at high altitudes above the most irradiated levels?), perhaps with mutations—hence, evolution—ensuing rapidly in future generations. And whether or not such a catastrophe happens, it seems plausible that if we emerged from Australopithicines in a few million years, then in a few million more years (10 million, if you want to be conservative) biological changes might have moved us to a level where we have little in common with our present species. Thus, instead of dying out, perhaps civilizations change so fast as to become virtually unrecognizable in less than a million years.

What seems more reasonable to me is to take into account biological and social evolution, and say that the time-slot in which we must encounter a life-bearing planet in order to meet creatures with some intelligence and technology, *who are evolutionarily close enough to us to want to communicate with us*, is longer than a thousand years but much less than half the life of the planet. Note that while a few of our scientists have "contacted" a few individuals or colonies among the chimpanzees, our species as a whole makes no great effort to maintain a continuing dialogue with the chimp species. Analogously, an alien species whom we might visualize as evolved 10 million years "beyond" us might have little interest in contacting us. This is contrary to the ethnocentric image of countless science-fiction films in which galactic civilizations send their ambassadors to us because we have reached the "maturity" of inventing the atomic bomb!

WHY HAVEN'T THEY CONTACTED US?

The upshot is that instead of assuming that the time-slot for finding kindred spirits on a planet is as long as half the life of the planet, as we assumed earlier, or as short as a thousand years, we should take an intermediate value. The best value might be the total lifetime of our species, perhaps a few million years, which is one-thousandth the lifetime of the planet. Applying this to our previous calculation, we might conclude that there are not 5 million civilizations, or only 10, but rather something closer to 10,000 civilizations in the galaxy today. The nearest one in this case would be roughly 1,000 light-years away.

Note that these are the figures for a technical civilization—communicative creatures with tools. On the basis of our initial discussion, we would infer that the number of evolved lifeforms—species analogous to Earth's trees, trilobites, zebras, slugs and Venus's-flytraps—is closer to the 50 million figure so that the nearest planet with well-evolved species might be within a few hundred light-years.

Quite a range of uncertainty: the nearest civilization being a few hundred light-years away according to one estimate; ten thousand in another; and perhaps 1,000 in our favored estimate! Such are the uncertainties and hazards of trying to be quantitative without exact data!

But what are the consequences if we're right? What if there *is* a civilization 1,000 light-years away from us? Could we make contact? If its creatures are more advanced than we are, have they colonized space, and if so, why haven't they reached us? What might they be like?

According to our best current understanding of the universe, nothing can travel faster than the speed of light. This means that if the nearest civilization is 1,000 light-years away, its best spaceships would take 1,000 years to reach us. Similarly, their radio messages would take 1,000 years to reach us. Some scientists have suggested that this may be the ultimate reason that our skies are not teeming with interstellar spaceships, nor our radio dials with interstellar messages. Civilized centers are just too far apart. Other scientists have countered with various speculative models. For example, a favorite science-fiction theme is the multi-generation interstellar spaceship. Even if nothing can travel faster than light, a civilization that has spread through its own planetary system could build a giant craft, or hollow out an asteroid to make living quarters, and launch it on an interstellar expedition that might take centuries, flying from one star to another in search of new homes. A vigorous civilization could easily send out hundreds of seed pods of this sort. Even if no system were readily found, generations of individuals could live happy lives in those self-sufficient giant pods (our own Earth is a self-sufficient spaceship). And when a planetary system was found, the population could expand and cultivate it. The colonies would be only five, ten or twenty light-years apart and could maintain radio contact during the early centuries of colonization. The new worlds, in turn, could eventually send out their own seed pods, and in this way the species could spread itself over a sizable part of the galaxy in astronomically short times. Though this scenario looks plausible, it has apparently not happened in our part of the galaxy, or else our own planetary system has been carefully avoided.

Perhaps other civilizations, realizing that the time needed to find neighbors by spaceship or to communicate with them by two-way radio is much longer than their own lifetimes, opt instead to send out radio signals to tell about themselves. You may say: "Why would anyone send out a signal if there's no hope of getting an answer back?" But we do it when we bury time capsules in shopping centers. And a writer does it when he writes a book that he hopes will be enjoyed by readers even after he's gone. So it's plausible that other civilizations might broadcast coded messages with information on themselves and how they perceive the universe. The broadcasts might last for as long as the

Pamela Lee

The peak sensitivity of the human eye is to yellowish light, the same color emitted by the sun as its maximum radiation. This suggests that lifeforms may evolve to be most sensitive to the wavelengths of light most strongly emitted by their parent star. Thus, creatures of a red star's or a blue star's world might perceive landscapes differently from the way we do. This landscape is on an imaginary planet of a low-mass red star, which is barely red-hot. The view at the top shows the scene as it might be perceived by a human observer, who would barely be sensitive to the dim, red light. The same scene (below) is shown as it would be perceived by the native creatures, whose eyes are sensitive to the infrared light most strongly emitted by the star; the star looks brighter, as do the red tones in the rocks.

Tom Miller

More than one intelligent species has evolved on this ice-bound planet, orbiting a sun on the outskirts of a globular cluster. An inhabitant of the volcanically heated valley is meeting with a highlander to trade his metal tools for the other's furs and fabrics. Cultural and physical differences are put aside because of their mutual dependence. Many different lifeforms could evolve intelligence independently on a world where pockets of evolution have been isolated by mountains or oceans. On Earth, the unique species that evolved on the island continent of Australia is a good example. Learning to cope with the adverse conditions of a hostile environment may be an incentive to the development of intelligence.

civilization itself or perhaps only intermittent messages might be sent. We ourselves beamed one such message toward a globular cluster in a somewhat tongue-in-cheek test by radio astronomers. And, of course, our weaker commercial radio and TV are already beaming their way out through the nearby stars; those signals are already tens of light-years away!

Radio astronomers in both the United States and the Soviet Union have already started some test programs of pointing radio telescopes toward various stars, "listening" for possible intelligent signals. So far, the results have been negative. Some scientists have argued philosophically that the effort is foolish, claiming success is so unlikely that we should use the money to support other projects. This seems to me shortsighted. As I have stressed, we *know* very little of what we're talking about here. We lack observations either positive *or* negative. For all we know, every hundredth solar-type star has a planet with a bustling civilization crying out to be heard! Conceivably, everybody else is tuned in to the galaxy's biggest computer-mail network, which we are only on the verge of discovering.

The assumption that some aliens might be broadcasting implies, of course, that the aliens have a psychology something like ours. This may be an ethnocentric mistake. We don't know that evolution has to produce curious, moderately aggressive, oversexed explorers like ourselves. The seas cover more of the Earth than the land and have had a longer history of biological evolution, but neither dolphins nor anybody else down there that we know of have produced hammers or radios. If civilizations produce radio builders only one time in a million, then we need another factor of one million in our chain of multipliers and the civilization we're looking for could be still farther away.

At any rate, going outdoors and looking is *always* better than sitting at home arguing. Using reasoning similar to that above, we would conclude that, at any one moment, weak signals might be reaching Earth from millions of sources more advanced than Earth (perhaps so far advanced that any messages in their radio signals would be indecipherable)! This would imply signals from one star system in 100,000 or so. Therefore, the search program would have to listen carefully, with great sensitivity, checking many wavelengths, to as many as 100,000 stars or more before we might expect success. Such a complete program is beyond our present means, but since our numbers may be far off the mark, the only way to improve our understanding is to make a start. This means funding a radio astronomy program to listen to and analyze signals. It would make a nice choice for an internationally funded program, an adventure for all humanity—looking to see who, if anyone, is out there in our cosmic environment. It would add to our understanding of the universe even if we got a negative result: to have studied a sample of 10,000 stars, for example, and established that there are no strong artificial signals.

We could go on and on, ever farther into the desert of data-unsupported speculation—the *terra prohibita* of scientists. It is usually less fruitful to pursue detail beyond the limit that can be supported by knowledge than to drop back and try to synthesize our supply of crude but valid information. So, let us summarize by saying that extrasolar planets seem very likely, and lifeforms on some of those also seem very likely. Therefore, intelligent lifeforms on at least a fraction of those planets also seem very likely, though we don't quite know whether "intelligent" means "like us." Imagine, as you stand under the night sky, that hidden in the sky like secret microdots with coded messages, nestled against some of those stars, are planets where waves break on coastlines; animals pursue their prey; skeptical thinkers ponder equations; creatures carry out acts of heroism or deceit or fear or love, by their own standards; and a youngster ponders the night sky and wonders if anyone is there.

A "cat's-eye eclipse." Here we imagine a view of an eclipsed moon of a hypothetical planet circling a rapidly rotating star, such as Pleione, in the star cluster known as the Pleiades. Evidence suggests that Pleione rotates 100 times faster than the sun, so rapidly that it is highly flattened, creating an elongated light source. The star is out of the picture, behind us, but casts a shadow of the planet on the moon. Depending on the relative distances, the planet may cast an elongated shadow, giving a curious appearance to the eclipsed moon.

Ron Miller

EPILOGUE: ENDLESS POSSIBILITIES

The possibilities of the universe seem endless. Among a thousand billion suns and a thousand billion worlds, what is inconceivable? To assess the plausible scenes in the universe one might simply fantasize stars, planets and creatures out of whole cloth, unfettered by current scientific knowledge. Perhaps much of what we can imagine can exist, and therefore *does* exist, somewhere. But there is a more orderly approach, which we have followed here. That is to start our imaginings with multitudes of individual physical realities and processes that we *know* to exist, and then combine them to visualize new situations, most of which almost certainly do exist somewhere.

For example, in this book, as well as in our companion volumes, *The Grand Tour* and *Out of the Cradle*, we have seen: blue stars, red stars, exploding stars, stars enveloped in dust clouds, stars as small as an asteroid, stars as big as Mars' orbit, star remnants that beep, star remnants that swallow light; red-glowing, green-glowing and blue-glowing gas clouds between the stars; star-disks that squirt out beams of glowing gas at a quarter the speed of light and galaxy disks a hundred million times bigger that do the same thing; worlds with oxygen/nitrogen air, worlds with carbon dioxide air, worlds with hydrogen/helium air, worlds with methane air, and many, many worlds with no air; worlds with wide rings and others with narrow rings; worlds with aurorae, dust storms, sulfuric acid rain, frozen smog swamps, nitrogen oceans, water oceans, water cress, trilobites, redwoods, bluejays, iguanas, lichens, ladybugs, sharks, bees, microbes living in cracks in rocks and

Ron Miller

worms inhabiting sulfur vents in ocean depths. We've seen big moons, little moons, moons of ice, moons of rock. Worlds whose climates change because of asteroid impacts every hundred million years or so; worlds smashed by impacts of giant asteroids, and new worlds re-created therefrom. All these we *know* to exist. We've *observed* them.

Starting with such a list, we can imagine all sorts of additional possibilities: a planet that started as an icy world, like Pluto, Uranus or Ganymede, but whose sun has turned into a red giant, melting the ice and creating a short-lived clement environment, with warm, fertile oceans. A world without seasons: its polar inclination is zero, its sun stays always over its equator, and the equator is too hot for liquid water, while only at the poles are warm sea waters. A massive world spinning rapidly because of the ancient off-center impact of a large asteroid; gravity is very stong at its poles, but very weak at its equator because of the "negative-g" effect of centrifugal force. (Such a world is not a new idea: a similar world was depicted by Hal Clement in his novel *Mission of Gravity*).

Everywhere we look throughout the sky there is a new marvel. Here is a galaxy whose nucleus has just "gone critical" in some sense we don't yet fully understand. Perhaps through explosive activity of a giant black hole's accretion disk, it has turned itself into a quasar and has blasted with X-rays and gamma rays a thousand worlds throughout its inner spiral arms, destroying uncounted species: struggling lichens, plankton, mosses, and salamanders that could talk. In another galaxy is a world with a small but close moon, which, due to tidal forces, has been spiraling inward, ever closer to the planet, and will crash into it in 6,000 years. The planet is inhabited and its creatures have begun to understand astronomy. Their scientists have just realized the catastrophe that will befall their civilization. An idea is dawning: is it possible to build vehicles to escape, reach other worlds in the system and colonize them?

An entire planetary system—a star with eight orbiting planets, one occupied—has been perturbed from its galactic orbit by the close passage of a massive globular cluster and flung out of its galaxy...out into the mist-free blackness of intergalactic space where the night sky has no stars at all, but only some fuzzy galaxy-patches of light. What

Jets from active galaxies may contain a flood of charged atomic particles. Here, such a jet passes through a close satellite galaxy containing the foreground planet, inducing massive auroras in the planet's atmosphere near the planet's two poles. This occurs in the same way that charged particles from solar flares cause "northern lights" and "southern lights" on Earth. Such radiation may preclude the evolution of life on this planet.

Ron Miller

will be the astronomy and mythology of these creatures, if intelligence and science appear among them?

If the volume of the universe is finite, then there must be a class of events whose probability of happening at all in that volume is less than one within the age of the universe. In other words, there must be some events so improbable that they have not yet happened. But the universe is so vast that the possibilities remain great. We are free to envision not only the class of "commonplace" events that must be happening somewhere now, but also the class of bizarre events that have happened only a few times, if ever.

In which class is the planet we know, with its oceans of liquid water, in which creatures evolved whose descendants walk balanced on their two hind legs and threaten each other with starfire bombs? We don't know yet whether this is one of the commonplace possibilities, or one of the bizarre possibilities of the universe. If we pursue the adventure of astronomical discovery, we have a good chance of finding out within our lifetime.

Perhaps somewhere else in the universe intelligent beings gaze into the sky and wonder if there might be strange creatures like ourselves living on some other planet orbiting one of the distant stars. Their curiosity might be great enough to inspire them to want to explore space, to build spaceships capable of visiting other worlds. If a thirst for knowledge is not unique to human beings, is it possible that one day a Columbus from some far-off world will set foot—or claw, or tentacle—on the unexplored shores of Earth?

GLOSSARY

A

accretion disk: a disk-shaped nebula of gas and/or dust surrounding a star or black hole, in which the gas or dust is slowly spiraling inward, eventually "accreting," or merging with the star's mass.

active galaxies: galaxies with nuclei that radiate more energy than do the nuclei of normal galaxies.

Alpha Centauri: the star system closest to Earth. Of its three co-orbiting stars, the closest one is too faint to see with the naked eye; the other two form a bright star prominent from Hawaii and latitudes farther south.

Andromeda galaxy: the closest spiral galaxy, similar in size and shape to our own galaxy. It is visible to the naked eye from the northern hemisphere as a fuzzy, luminous patch in the constellation of Andromeda.

astrometry: the study of positions and movements of celestial bodies.

B

Barnard's Star: the second closest star system, once regarded as having planet-like companions of about Jupiter's mass. Recent data indicate that the earlier "detection" of these planet-like objects was incorrect.

Beta Pictoris: a hydrogen-burning star recently found to have a large, disk-shaped cloud of dust around it. It is regarded as possible evidence of planetary material around other stars.

Big Bang: the explosion-like event, roughly 16 billion years ago, in which all mass in the known universe began expanding from a fireball of extremely high temperature and density.

binary star: a pair of stars orbiting around each other, or, more precisely, around their common center of mass.

bipolar outflow: the high-speed jetting of gas in opposite directions from disk-shaped systems, usually in directions perpendicular to the plane of the disk. The mechanism is poorly understood. Bipolar jets have been seen both from disks around stars and from galaxies.

black hole: an extremely dense object whose surface gravity is so intense that material could escape only if it were propelled to a speed faster than the speed of light. Because nothing can move faster than light, mass and light cannot normally escape from a black hole. Black holes may form from some supernova explosions; these probably have masses greater than about six solar masses.

brown dwarf: a substellar object.

C

cocoon nebula: an opaque cloud of dust and gas surrounding a protostar or newly formed star.

collapse: to contract rapidly, relative to astronomical scales. Self-gravitating, shrinking nebulae are said to collapse into stars; the process may last some thousands of years.

colliding galaxies: a subclass of radio galaxies involving pairs of galaxies colliding with each other.

contact binary: a binary star system in which the two members touch. Usually, the two members are red giants and touch only after they expand to the red-giant state.

Copernican revolution: the discovery that the Earth is not at the center of the planetary system but rather that the planets move around the sun.

Doppler effect: the shift in wavelength of radiation from an object, as perceived by an observer, if the object is moving relative to the observer. The amount of Doppler effect can be used as a measure of the rate of motion toward or away from the observer.

double star: a binary star.

dwarf elliptical galaxy: the most common kind of galaxy, a sort of glorified globular cluster.

E

elliptical galaxy: a galaxy of ellipsoidal form, usually shaped somewhat like a flattened orange or football.

emission nebula: a nebula shining because of excitation of its gas by the radiation from a nearby star. The gas, once excited, emits light of its own.

escape velocity: at a specified distance from an object, the velocity to which a small body would have to be accelerated in order to keep moving outward into space and not fall back onto the larger body; the small body is thus said to "escape" from the gravity of the larger one.

event horizon: the closest distance to a black hole from which matter or light can escape.

F

Flammarion effect: a term coined to describe the multicolor shadows found in binary star systems whose parent stars have different colors. Named after Camille Flammarion, the French author who first described it in the late 1800s.

G

galaxy: the largest systems of stars, often containing 100 billion stars.

giant elliptical galaxy: the largest of the elliptical galaxies, which include the largest and most massive galaxies known.

globular cluster: a type of densely packed, spheroidal star cluster, often with 100 thousand or a million stars. They tend to be located in swarms surrounding galaxies.

gravity: the force of attraction between all masses.

H

helium: the second most abundant element in the universe. A helium atom consists of a nucleus with two protons and two neutrons, circled by two electrons.

Hubble revolution: a term coined by us, in parallel with "Copernican revolution," to denote the discovery that our Milky Way galaxy is not unique. American astronomer Edwin P. Hubble discovered in the 1920s that other galaxies are composed of stars, like our own. Prior to that time, galaxies were thought to be a variety of nebulae within our Milky Way.

hydrogen: the lightest, simplest and most abundant element in the universe. A hydrogen atom consists of one proton in the center, circled by one electron. Hydrogen is a very lightweight gas, found in only small amounts in Earth's atmosphere.

hydrogen-burning star: the most common kind of star. Stars spend most of their lifetime "burning" hydrogen, during which they have a relatively constant appearance. A star with the mass of the sun spends around ten billion years in this stage; more massive stars spend less time.

I

infrared heat radiation: infrared light emitted by a body due to its own heat. The amount of such radiation can be used to measure the temperature of the body. Planets, dust around stars and very cool stars radiate large amounts of infrared heat radiation, which is invisible to the eye but can be detected by astronomers' instruments on large telescopes.

infrared light: light of a wavelength too long (too red) to be detected by the human eye.

Jupiter mass: the amount of mass in the planet Jupiter; a convenient measure for the masses of substellar objects and planets.

Magellanic Clouds: two small, irregular galaxies that are satellites of the Milky Way galaxy; on Earth, they are visible only from the equatorial regions and southern hemisphere.

Milky Way: the band of light across our night sky, produced by the light of innumerable stars too far to be resolved separately as we look out through the plane of our galaxy's disk.

molecular cloud: an unusually dense and large interstellar nebula of gas and dust. Because of the high density of gas, atoms combine into molecules with more than average frequency. Regions of star formation are usually located near or within molecular clouds.

multiple star (or multiple-star system): three or more stars orbiting around each other.

nebula: a cloud of gas and dust in interstellar space or around a star.

neutron star: an extremely dense form of star, on the order of a mile in diameter, produced in the final evolutionary stages of some stars more massive than the sun. Neutron stars form as the result of some supernova explosions and probably have masses in the range of about 1.4 to 6 solar masses.

nuclear reaction: reaction involving the collision of atomic nuclei.

nucleus (of atom): the central particle in an atom, composed of protons and neutrons.

nucleus (of galaxy): the central, brightest object in a galaxy.

photon: a unit of light, visualized as having some properties of a tiny, massless particle smaller than an atom.

planetary nebula: a nebula of roughly spheroidal shape, usually ejected from a star located at its center. Because of its round shape, it may resemble the disk of a planet as seen in a telescope, but there is no physical connection with planets.

planetesimal: a solid body, growing in a cocoon nebula; a precursor of a full-fledged planet. The term is generally used to denote a rocky and/or icy body with dimensions ranging from millimeters to hundreds of kilometers, but smaller than a planet.

population I: stars similar in composition to the sun. They tend to be located in spiral arms of spiral galaxies.

population II: stars much poorer in heavy elements than the sun; they are almost pure hydrogen and helium. They tend to be located in elliptical galaxies, central regions of spiral galaxies and globular clusters.

protostar: a mass of material in the process of forming a star but before onset of nuclear reactions.

pulsar: a neutron star that appears to flash rapidly, due to effects associated with its extremely rapid rotation.

quasar: a very highly red-shifted object of extremely high intrinsic luminosity. Quasars are thought to be highly energetic nuclei of galaxies, probably a form associated more commonly in galaxies' youth than in their old age. The most distant, oldest, most highly red-shifted objects we can see are quasars.

R

radio galaxy: a galaxy that gives off natural radio waves.

radio telescope: a giant radio antenna designed to concentrate radio waves and allow detection of faint radio signals reaching us from distant nebulae and galaxies.

red giant: a huge form of star, often a hundred times as large as the sun, produced only as a temporary stage near the end of a star's evolution. A red giant has a vast, expanded, low-density atmosphere and looks redder than most stars because its atmosphere is cooler than that of most stars.

red shift: a Doppler shift toward longer (redder) wavelength.

reflection nebula: a nebula shining by reflected starlight.

resolve: to distinguish individually. Two stars, close to each other, are said to be resolved if a telescope can show them as two separate objects rather than blurred together into one.

S

Sagittarius: the constellation of the Archer; it is important because its region of the sky is the direction in which the nucleus of our galaxy is located.

Shapley revolution: a term coined, in parallel with "Copernican revolution," to denote the discovery by American astronomer Harlow Shapley that Earth and its sun are not at the center of our galaxy.

solar mass: the amount of mass in the sun.

solar system: the sun, its planets and the smaller bodies orbiting the sun, including asteroids and comets.

spiral galaxy: a disk-shaped galaxy in which spiral arms can be traced.

star: a spheroidal mass of gas large enough to initiate nuclear reactions in its center. Minimum mass is about 0.085 solar mass, or about 85 Jupiter masses.

starquake: a hypothetical earthquake-like movement or fracturing in the crust of a neutron star.

stellar wind: outward flow of gas from a star.

substellar object: a starlike object too small to initiate nuclear reations in its core; in practice, substellar objects have masses between about 2 and 85 Jupiter masses.

supernova: an explosion of a massive star that becomes unstable. Stars greater than roughly six solar masses may evolve so fast that they become unstable and explode to form a supernova.

surface brightness: the amount of light radiated per unit angular area, as seen by an observer; for instance, the amount of light per square degree from a galaxy.

T

total apparent brightness: the total amount of light reaching a given observer from an object.

U

universe: everything that exists.

W

white dwarf: a very small, hot form of star, reached at the end of most stars' evolution. Its whitish or blue-white appearance is caused by surface temperatures similar to or hotter than those of the sun.

INDEX